SpringerBriefs in Applied Sciences and Technology

Nonlinear Circuits

Series Editors

Luigi Fortuna, DIEEI, Università di Catania, Catania, Italy
Guanrong Chen, Kowloon, Hong Kong SAR, P.R. China

SpringerBriefs in Nonlinear Circuits promotes and expedites the dissemination of substantive new research results, state-of-the-art subject reviews and tutorial overviews in nonlinear circuits theory, design, and implementation with particular emphasis on innovative applications and devices. The subject focus is on nonlinear technology and nonlinear electronics engineering. These concise summaries of 50–125 pages will include cutting-edge research, analytical methods, advanced modelling techniques and practical applications. Coverage will extend to all theoretical and applied aspects of the field, including nonlinear electronic circuit dynamics from modelling and design to their implementation. Topics include but are not limited to:

- nonlinear electronic circuits dynamics;
- oscillators;
- cellular nonlinear networks;
- arrays of nonlinear circuits;
- chaotic circuits;
- system bifurcation;
- chaos control;
- active use of chaos;
- nonlinear electronic devices;
- memristors;
- circuit for nonlinear signal processing;
- wave generation and shaping;
- nonlinear actuators;
- nonlinear sensors;
- power electronic circuits;
- nonlinear circuits in motion control;
- nonlinear active vibrations;
- educational experiences in nonlinear circuits;
- nonlinear materials for nonlinear circuits; and
- nonlinear electronic instrumentation.

Contributions to the series can be made by submitting a proposal to the responsible Springer contact, Oliver Jackson (oliver.jackson@springer.com) or one of the Academic Series Editors, Professor Luigi Fortuna (luigi.fortuna@dieei.unict.it) and Professor Guanrong Chen (eegchen@cityu.edu.hk).

Members of the Editorial Board:

Majid Ahmadi; Wayne Arter; Adi Bulsara; Arturo Buscarino; Syamal K. Dana; Mario Di Bernardo; Alexander Fradkov; Mattia Frasca; Liviu Goras; Mo Jamshidi; Mario Lavorgna; Shin'ichi Oishi; Julien C. Sprott; Alberto Tesi; Ronald Tetzlaff; Mustak E. Yalcin; Simin Yu; Jacek M. Zurada

Publishing Ethics: Researchers should conduct their research from research proposal to publication in line with best practices and codes of conduct of relevant professional bodies and/or national and international regulatory bodies. For more details on individual ethics matters please see: https://www.springer.com/gp/authors-editors/journal-author/journal-author-helpdesk/publishing-ethics/14214

More information about this series at http://www.springer.com/series/15574

Müştak E. Yalçın · Tuba Ayhan ·
Ramazan Yeniçeri

Reconfigurable Cellular Neural Networks and Their Applications

 Springer

Müştak E. Yalçın
Department of Electronics and
Telecommunications Engineering
Istanbul Technical University
Istanbul, Turkey

Tuba Ayhan
Department of Electronics and
Telecommunications Engineering
Istanbul Technical University
Istanbul, Turkey

Ramazan Yeniçeri
Aeronautical Engineering
Istanbul Technical University
Istanbul, Turkey

ISSN 2191-530X   ISSN 2191-5318   (electronic)
SpringerBriefs in Applied Sciences and Technology
ISSN 2520-1433   ISSN 2520-1441   (electronic)
SpringerBriefs in Nonlinear Circuits
ISBN 978-3-030-17839-0   ISBN 978-3-030-17840-6   (eBook)
https://doi.org/10.1007/978-3-030-17840-6

This Springer imprint is published by the registered company Springer Nature Switzerland AG
The registered company address is: Gewerbestrasse 11, 6330 Cham, Switzerland

Contents

Chapter 1
Introduction

1.1 Why This Book?

Conventional algorithmic solution for today's engineering problems is started to digitize the sensory data and then process this raw data on a conventional computer architecture. To obtain real-time response from the algorithms, low latency is required which demands to process huge amount of input data. When the biological sensing systems which can handle the same tasks in real time are considered to mimic nature, they are able to complete processing tasks to extract salient information from the incoming sensory data, thus eliminating the redundant data before transmitting them for subsequent processing for analyzing and making decision.

Cellular Neural Network (CNN) [1] which is inspired by architecture of biological networks is reasonable to be used in sensory data processing to extract salient information. Furthermore, its conversion to a many-core microprocessor which is named as CNN-Universal Machine (CNN-UM) [2] can handle arrays of continuous-time dynamic signals as variables and applies sequences of spatial–temporal dynamic operators and logic functions as instructions make practical for feature extraction process. Since its first analog realization [3], CNN became very approved in visual processing. Today, mixed-signal integration of CNNs enables direct interface to sensory device such as ACE16k [4], which is specially designed for image processing applications. Fusing the sensory and the processing circuitry on the same chip helps to overcome the drawbacks of real-time signal processing of conventional computer architecture.

CNN traditionally includes only one type of neuron. That limits the CNN with one neuron population. However, collaboration of subpopulations is the key concept in processing and decision mechanisms of biological sensing systems. Furthermore, the connection topology of CNNs is completely regular and it is not allowed to randomness. Two key concepts of making bioinspired sensory data processors are to consider collaborative working of neuron subpopulations and build the connections in halfway between fully random and regular. In this book, reconfigurable cellular neural network which is enhanced by collaboration of different neuron populations is

M. E. Yalçın et al., *Reconfigurable Cellular Neural Networks and Their Applications*, SpringerBriefs in Nonlinear Circuits, https://doi.org/10.1007/978-3-030-17840-6_1

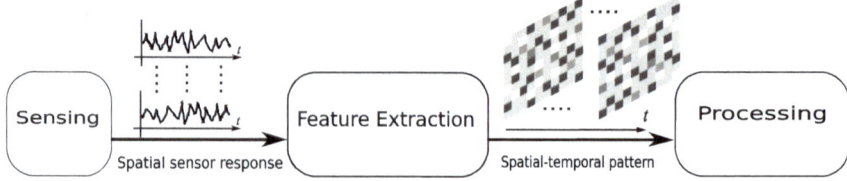

Fig. 1.1 A block diagram of biological signal processing system which has three main parts: sensing, feature extraction, and processing. Feature extraction subblock maps spatial sensor response into spatiotemporal patterns. Processing subblock produces a decision using these spatiotemporal patterns as an input. This book introduces CNN-based network models to handle feature extraction subblock together with their implementations and applications

used in bioinspired applications while protecting the simplicity of its implementation. Figure 1.1 shows the basic subblocks of a biological signal processing system. This book introduces CNN-based network models to handle feature extraction subblock converting spatial sensor response into spatiotemporal patterns.

We start with introducing Wilson–Cowan neuron population in Sect. 2.2. Model describes the evolution of two types of neuron activity in a synaptically coupled neuronal network. Inspired by Wilson–Cowan model, CNN models which include collaborative working of neuron subpopulations with a connection topology halfway between fully random and regular connections are presented in Sect. 2.4 after CNNs' overview in Sect. 2.3. We also present a CNN model in Sect. 2.3.2 which is named locally coupled oscillatory network to generate active spatiotemporal waves. Chapter 2 fully address the Artificial Neural Network (ANN) models that are employed in the book.

CNN's main application area is early image processing because of its regular two-dimensional array structure. In Chap. 3, we will exploit the ANN models which are utilized by allowing subpopulations and random connections for odor processing applications. Also, the networks are tested with the real measurements in this chapter to justify their performances. In addition to these tests, the effect of network topology on feature extractor's performance is investigated on reconfigurable cellular neural network reconfiguring its topology with randomly build new topologies.

Besides their computational capabilities and energy efficiencies of biological systems are always inspiring for engineers, their implementation using the current technology must be practical and feasible. In Chap. 4, the proposed network models in Chap. 3 are implemented on Field-Programmable Gate Arrays (FPGAs) which are state-of-the-art reprogrammable integrated circuits. From embedded system design perspective, custom single-purpose processors are designed to demonstrate the effectiveness of the models. Furthermore, the designs are not only given for CNN subparts, but full system designs to handle the applications which are odor processing and motion planning are considered in the chapter.

This book systematically gathers author's research activities on cellular neural network which have been published in international journals, conferences, and workshops [5–14]. Gathering research papers have been the concerned applications and neural network models based on CNNs which are modified to be used in these applications and implementations of the CNN's model on FPGAs. Furthermore, the book is supplied by the essential background information for facilitating a better understanding.

References

1. L.O. Chua, L. Yang, Cellular neural networks: theory and applications. IEEE Trans. Circuits Syst. I **35**(10), 1257–1290 (1988)
2. T. Roska, L. Chua, The CNN universal machine—an analogic array computer. IEEE Trans. Circuits Syst. II Analog Digit. Signal Process. **40**(3), 163–173 (1993)
3. S. Espejo, C. Carmona, R. Dominguez-Castro, A. Rodriguez-Vazquez, CNN Universal chip in CMOS technology. Int. J. Circuit Theory Appl. **24**, 93–111 (1996)
4. A. Rodriguez-Vazquez, G. Linan-Cembrano, L. Carranza, E. Roca-Moreno, R. Carmona-Galan, F. Jimenez-Garrido, R. Dominguez-Castro, S. Meana, ACE16k: the third generation of mixed-signal SIMD-CNN ACE chips toward VSoCs. IEEE Trans. Circuits Syst. I Regul. Pap. **51**(5), 851–863 (2004)
5. R. Yeniceri, M.E. Yalcin, An emulated digital wave computer core implementation, in *European Conference on Circuit Theory and Design, ECCTD 2009* (2009), pp. 831–834
6. R. Yeniceri, M.E. Yalcin, Path planning on cellular nonlinear network using active wave computing technique, in *Proceedings of SPIE, Bio-engineered and Bioinspired Systems IV*, vol. 7365 (2009)
7. T. Ayhan, K. Muezzinoglu, M.E. Yalcin, Cellular neural network based artificial antennal lobe, in *Proceedings of the 12th IEEE International Workshop on Cellular Neural Networks and Their Applications (CNNA 2010)* (2010), pp. 1–6
8. V. Kilic, R. Yeniceri, M.E. Yalcin, A new active wave computing based real time mobile robot navigation algorithm for dynamic environment, in *12th International Workshop on Cellular Nanoscale Networks and Their Applications (CNNA)* (2010), pp. 1–6
9. T. Ayhan, Using CNN baased antennal lobe model to accelerate odor classification. ITU, Graduate School Of Science, Engineering and Technology, M.Sc. Thesis, Istanbul, October 2010
10. T. Ayhan, M.E. Yalcin, Randomly reconfigurable cellular neural network, in *Proceedings of the 20th European Conference on Circuit Theory and Design (ECCTD11)* (2011), pp. 625–628
11. T. Ayhan, R. Yeniceri, S. Ergunay, M.E. Yalcin, Hybrid processor population for odor processing, in *2012 IEEE International Symposium on Circuits and Systems (ISCAS)* (2012), pp. 177–180
12. R. Yeniceri, M.E. Yalcin, A new CNN based path planning algorithm improved by the Doppler Effect, in *13th International Workshop on Cellular Nanoscale Networks and Their Applications (CNNA)* (2012), pp. 1–5
13. T. Ayhan, M.E. Yalcin, An application of small-world cellular neural networks on odor classification. Int. J. Bifurc. Chaos **22**(1), 1–12 (2012)
14. R. Yeniceri, Implementations of novel cellular nonlinear and cellular logic networks and their applications. ITU, Graduate School Of Science, Engineering And Technology, Doctorate Thesis, Istanbul, October 2015

Chapter 2
Artificial Neural Network Models

2.1 Neural Network Models

Modeling neural networks have always been an interesting question for various research and application areas. The reason behind that is not questionable: brain evolution is one of the biggest mysteries for us, and computational capabilities of the brain have always been inspiring for engineers. Biological neurons inspired Rosenblatt for the initial formal neuron model, perceptron [1]. On the other hand, the layered structure of biological neural networks leads the field of Artificial Neural Networks (ANN). Feedforward ANNs are developed into recurrent ANNs and recursive ANNs with the inclusion of feedback loops. Moreover, the connection structures are diversified into networks like Hopfield [2] and Cellular Neural Networks (CNNs) [3]. The connections between neurons have been modeling using Electroencephalography (EEG) and Magnetoencephalography (MEG) signals. These signals reveal that the complexity of these connections makes difficult to understand neurons individually.

Neural mass model which was introduced by Wilson and Cowan [4] offers an alternative approach to understand the behavior of biological neural networks. Neural mass model is based on the collaboration of excitatory and inhibitory neuron populations. This approach is more useful than modeling the neurons individually because the interactions between neurons make a neural area complicated to model. In addition, neural modeling approach is not suited to an investigation of those parts which are associated with higher functions such as sensory information processing and the attendant complexities of learning, memory storage, and pattern recognition [4]. Hartline and Ratliff are the first ones who showed the fact that all nervous processes of any complexity are dependent upon the interaction of excitatory and inhibitory cells [5]. The leading work of Wilson and Cowan proposes an equation set for neuron population and gives the results of computer simulations showing the interesting dynamics of their model. This attempt is efficient to mimic the EEG signals created by certain fields of the brain. Silva et al. [6] followed the previous attempt and proposed a model that generates alpha-like rhythm when driven by Gaussian noise. Freeman [7] provided compatible results with these models. Modeling

© The Author(s), under exclusive licence to Springer Nature Switzerland AG 2020
M. E. Yalçın et al., *Reconfigurable Cellular Neural Networks and Their Applications*,
SpringerBriefs in Nonlinear Circuits, https://doi.org/10.1007/978-3-030-17840-6_2

brain activity by means of neural population activity provides two major advantages. One is about mimicking the brain reactions such as sensory data processing without fully understanding the complex dynamics going on behind this process. Various types of sensory data from visual [5] to olfactory [8, 9] are processed with neuron populations. The second advantage that comes with using neuron population model-based artificial systems is that loss of an individual neuron/cell does not affect the process done in the artificial model much, and a significant damage in the output is not expected. Therefore, for realizations of bioinspired systems, modeling the system as an action of populations is more beneficial than focusing on individual processors. Ordinarily, the connection topology is assumed to be either completely regular or completely random, but many biological, technological, and social networks lie somewhere between these two extremes [10]. Therefore, fully deterministic networks are insufficient for mimicking those biological networks. However, most of the above neural fields are closely related to randomly connected nets [11]. An effective mimesis should represent both the randomness and deterministic characters of a biological network.

2.2 Wilson–Cowan Neuron Population Model

One of the most popular neuron population models, which includes excitatory and inhibitory interactions, is Wilson–Cowan population model. In the original work of Wilson and Cowan, $E(t)$ and $I(t)$ refer to activities in respective subpopulation and these functions are equal to the proportion of cells which are sensitive and receive at least threshold excitation at time t [4]. The expression for the excitatory subpopulation is given as

$$E(t+r) = \left[1 - \int_{t-r}^{t} E(\tau)d\tau\right]\mathscr{S}^{E}\{z\}, \qquad (2.1)$$

where r is the absolute refractory period and $\mathscr{S}(z)$ is the subpopulation response function, and the term $[1 - \int_{t-r}^{t} E(\tau)d\tau]$ shows the proportion of excitatory cells which are sensitive. The subpopulation response function is a function of the average levels of excitation within the subpopulations. The average level of excitation generated in an excitatory cell at time t is

$$z = \int_{-\infty}^{r} \alpha(t-\tau)[c_1 E(\tau) - c_2 I(\tau) - P(\tau)]d\tau, \qquad (2.2)$$

where $P(\tau)$ is the external input to the excitatory subpopulation, and c_1 and c_2 (both positive) are the connectivity coefficients. Similar formulation is valid for inhibitory subpopulation $I(t)$,

$$I(t+r) = \left[1 - \int_{t-r}^{t} I(\tau)d\tau\right]\mathscr{S}^I\left\{\int_{-\infty}^{r} \alpha(t-\tau)[c_3 E(\tau) - c_4 I(\tau)\right.$$

$$\left. - Q(\tau)]d\tau\right\}. \tag{2.3}$$

Using coarse-grained approximation [4], Eqs. (2.1) and (2.3) are rewritten as

$$\tau^E \frac{dE}{dt} = -E + (1 - rE)\mathscr{S}^E\{k^E c_1 E - k^E c_2 I + k^E P(t)\}$$

$$\tau^I \frac{dI}{dt} = -E + (1 - rI)\mathscr{S}^I\{k^I c_2 E - k^I c_4 I + k^I Q(t)\}. \tag{2.4}$$

Superscripts E and I stand for excitatory and inhibitory neurons, respectively. $(1 - rE)$ and $(1 - rI)$ show probability of sensitivities, and r and k are constants. $P(t)$ and $Q(t)$ are external inputs received by excitatory and inhibitory subpopulations, respectively.

If we look closely to the individual sensitive cells in the subpopulations, where x_i is the state variable of ith neuron in the excitatory subpopulation and y_i belongs to the inhibitory one [8],

$$\beta_i^E \frac{dx_i(t)}{dt} = -x_i(t) + K_i^E \theta^E\left(\sum w_{ij}^{EE} x_j(t) - \sum w_{ij}^{EI} y_j(t) + g_{inp}^E U_i^E\right) + \mu_E$$

$$i \in \{1, 2, ..., N_E\} \tag{2.5}$$

$$\beta_i^I \frac{dy_i(t)}{dt} = -y_i(t) + K_i^I \theta^I\left(\sum w_{ij}^{IE} x_j(t) - \sum w_{ij}^{II} y_j(t) + g_{inp}^I U_i^I\right) + \mu_I$$

$$i \in \{1, 2, ..., N_I\}. \tag{2.6}$$

Each neuron may have custom time constant β_i, excitation level K_i, and activation function θ. The weights denoted by w determine the connection between the cells. A cell from population F is connected to a cell of the population T with weight w_{ij}^{FT}. If the weight is zero, it means there is no direct connection between those cells.

In many biological sensory organs like olfactory brain, there is no connection between the neurons belonging to same subpopulation [12]. This also prevents saturation with excitation or inhibition. When this adapted model is rewritten for implementation, the states of the neurons are

$$\dot{x}_i = f\left(x_i, \mathfrak{I}_i^E, \mu_E\right)$$

$$\dot{y}_i = f\left(y_i, \mathfrak{I}_i^I, \mu_I\right), \tag{2.7}$$

where μ_E and μ_I are the system noises for excitatory and inhibitory neurons, respectively. $\mathfrak{I}^E$ and $\mathfrak{I}^I$ denote the activation level for corresponding neuron:

Fig. 2.1 Randomly located
six exhibitory (white) and
three inhibitory (black)
neurons. Possible
connections between the
neurons are illustrated in
figure

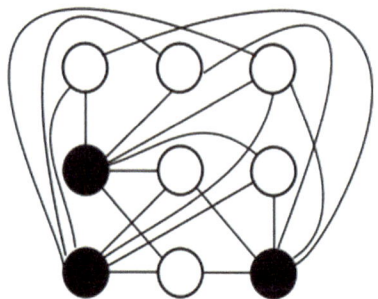

$$\mathfrak{J}_i^E = K_i^E \theta^E \left(-\sum w_{ij}^{EI} y_j(t) + g_{inp}^E U_i^E \right)$$
$$\mathfrak{J}_i^I = K_i^I \theta^I \left(\sum w_{ij}^{IE} x_j(t) + g_{inp}^I U_i^I \right). \tag{2.8}$$

Implementation of Wilson–Cowan-based model requires two unit processors exhibiting Eq. (2.7). As in biological structure, exhibitory and inhibitory cells are randomly located in the network and there is no connection between the exhibitory and inhibitory neurons. External input (sensor response) of a projection neuron is excited by the neuron, but in order to prevent this excitation go infinity, the sensor response is decremented by inhibitory neurons connected to that projection neuron. The collaboration of inhibition and excitation is the key concept to process information in neuron population models.

Figure 2.1 shows a neuron population which consists of six exhibitory ($N_E = 6$) and three inhibitory ($N_I = 3$) neurons. The neurons are randomly located, and there is no connection between the neurons belonging to the same subpopulation.

2.3 Cellular Neural/Nonlinear Network (CNN)

Cellular neural network [3] is an information processing system composed of a large number of simple analog signal processing elements, called cell, which are spatially arranged and locally coupled to perform parallel processing in order to solve a given computational task. The key concept that distinguishes a CNN from the other neural networks is to have local interconnection among its cells. Cellular neural network was renamed as cellular nonlinear network by Chua and Roska in 1993 with the same abbreviation. Cellular nonlinear network [13] defines locally coupled cells, where each cell is a dynamical system and it is not necessary that all the cells have the same dynamic. Cellular nonlinear network might consist of locally coupled many different dynamical systems which is used to process information and to generate spatiotemporal dynamic behavior.

1-D and 2-D CNNs are spatial arrangement of the cells in one- and two-dimensional arrays. A cell in 1-D and 2-D CNN is denoted by C_i where $i \in$

$\{1, 2, ..., M\}$ and $C_{i,j}$ where $i \in \{1, 2, ..., M\}$ and $j \in \{1, 2, ..., N\}$, respectively. Each cell in CNN is a dynamical system which has an input U, an output y, and a state x evolving according to some prescribed dynamical laws. The ith cell in a row of N cells is defined by the state equation

$$\dot{x}_i = g(x_i, z_i, \Im_i, U_i)$$
$$y_i = f(x_i), \tag{2.9}$$

where z_i is the threshold (usually assumed to be a constant), $\Im_i$ is a synaptic law of the cell, $f(\cdot)$ is the nonlinear output function, and $g(\cdot)$ is a function that defines the cell dynamic which might be any function [14].

The coupling between the cell C_i and its local cells (C_k) which are in *Sphere of Influence* of radius r:

$$S_i(r) = \{C_k : max(|k - i| \le r, \ 1 \le k \le N, \ k \ne i\} \tag{2.10}$$

is defined by the synaptic law. Figure 2.2 shows $S_i(1)$ and $S_i(2)$ on a row of seven cells.

The synaptic law of the cell is given by

$$\Im_i = \sum_{S_i(r)} \{A_{k-i} y_k + B_{k-i} U_i\}, \tag{2.11}$$

where A is the *feedback template* and B is the *feedforward template*. Templates which are also called synaptic matrices are space-invariant weights. Local coupling and to have invariant weight are indeed a great advantage for current planar Very-Large-Scale Integration (VLSI) technology [15].

One of the pioneering works in CNNs study is the CNN Universal Machine [16] which is the first algorithmically programmable analog processor introduced to the engineering community. Instructions which need to be executed by CNN-UM are represented by *cloning template* which consists of A, B, and threshold z.

In fact, spatial arrangement of processing cells in a two-dimensional array is very suitable for image processing such that each pixel of an image size of $N \times M$ matches cell's input $U_{i,j}$ or the state variable $x_{i,j}$ on corresponding 2-D CNN. Today, mixed-signal integration of CNNs enables direct interface to sensory device such as

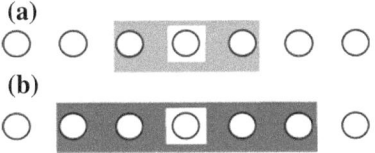

Fig. 2.2 Seven cells which are spatially arranged in 1-D CNN architecture. Sphere of influence $S_i(r)$ of radius **a** $r = 1$ and **b** $r = 2$ on this 1-D CNN is shown with dark regions

ACE16k [17], which is specially designed for image processing applications. Fusing the sensory and the processing circuitry on the same chip is helped to overcome the drawbacks of real-time signal processing of traditional digital computer.

2.3.1 2-D Cellular Neural Networks

A cell $C_{i,j}$ in a two-dimensional array is governed by

$$\dot{x}_{i,j} = g(x_{i,j}, z, \Im_{i,j}, U_{i,j})$$
$$y_{i,j} = f(x_{i,j}), \tag{2.12}$$

and its sphere of influence is defined by

$$S_{i,j}(r) = \{C_{k,l} : max(|k - i|, |l - j| \le r, \ 1 \le k \le N, \ 1 \le l \le M, \ k \ne i, l \ne j)\}. \tag{2.13}$$

The synaptic law of the corresponding cell is given by

$$\Im_{i,j} = \sum_{S_{i,j}(r)} \{\hat{A}_{k-i,l-j} y_{k,l} + B_{k-i,l-j} U_{k,l}\}. \tag{2.14}$$

Figure 2.3 shows spatially arranged cells in two-dimensional array, and sphere of influence of a cell for radius $r = 1$ on in this network is shown. In this case, feedback template $\hat{A}$ which defines the weights of the neighbor cell's outputs in the synaptic law is a 3×3 matrix. Furthermore, feedforward template B is also 3×3 matrix.

The interconnections among cells in the sphere of influence could be defined by the weights of the neighbor cell's states instead of the outputs. In this case, the synaptic law is given as

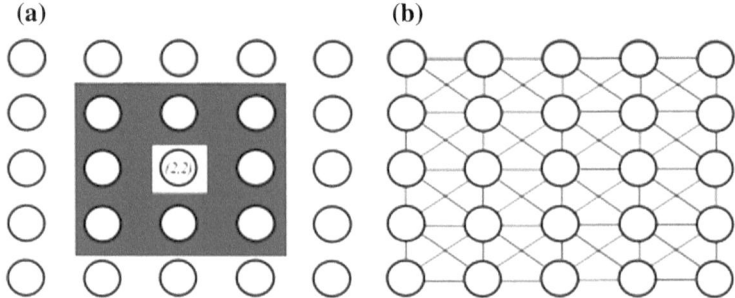

(a)   **(b)**

Fig. 2.3 Spatially arranged 30 cells in 2-D CNN architecture. **a** The neighboring cells ($C_{2,2}$, $C_{2,3}$, $C_{2,4}$, $C_{3,2}$, $C_{3,4}$, $C_{4,2}$, $C_{4,3}$, $C_{4,4}$) which lies inside the sphere of influence $S_{3,3}(1)$ are shown with dark regions, and corresponding local couplings between cells are shown in figure (**b**)

$$\Im_{i,j} = \sum_{S_{i,j}(r)} \{A_{k-i,l-j}x_{k,l} + B_{k-i,l-j}U_{k,l}\}, \tag{2.15}$$

where the weight A is named state-feedback template. Chua-Yang CNN [3] cell model which is the first and a very well-known cell model in the literature is given by

$$\dot{x}_{i,j} = -x_{i,j} + \Im_{i,j} + z$$
$$y_{i,j} = f(x_{i,j}) = \frac{1}{2}(|x_{i,j} + 1| - |x_{i,j} + 1|). \tag{2.16}$$

Sphere of influence (2.13) of a cell in Chua-Yang model is defined for radius $r = 1$, and the synaptic law $\Im_{i,j}$ is as given in Eq. (2.14). For the emulation purposes, discretized versions of CNN models have also been introduced. Basically, forward Euler integration method is used to obtain and discretized versions of CNN architecture are called discrete-time CNN (DT-CNN).

Cloning templates for Chua-Yang CNN model have been widely explored for image processing applications and they are collected in cellular wave computing library [18]. Edge detection template [18] is tested on 64×64 array of CNN using Fig. 2.4a as the input image. The test result is shown in Fig. 2.4b; for integration, time step is 0.1, initial state is full zero image, and the number of iteration is 100.

It has been reported that templates from the library were tested on the implemented on VLSI implementation of CNNs such as ACE4k and ACE16k; many of these templates were found to work incorrectly [19]. The reasons for that are erroneous chip behaviors. In [20], authors use measurements from actual chip as part of cost function for a global optimization method to find an optimal template. Most recently, a system-on-chip implementation which consists of a CNN emulator design and a processor which performs template learning algorithm has been presented [21]. In that work, templates can be updated by a learning algorithm in run time while unavoidable and undesirable behaviors exist.

CNN models include only one type of neuron, and it is created on a regular grid with a feedback template A that limits the CNN with one neuron population. However, collaboration of two or more subpopulations is the key concept in processing and

Fig. 2.4 Result when edge detection template [18] is tested on (**a**) 64×64 Lena image is given in figure (**b**)

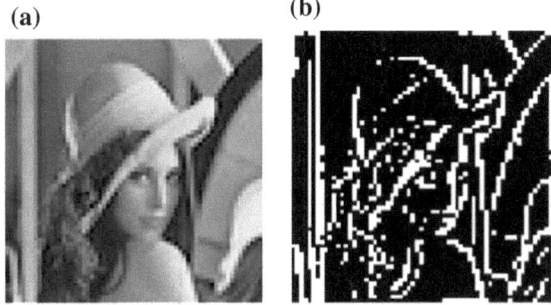

(**a**)   (**b**)

decision mechanisms of the human brain. The background served by neural mass models and realizability of CNN should be associated to improve the resemblance between CNN and nature.

2.3.2 Locally Coupled Oscillatory Network

Locally coupled oscillatory network is locally coupled oscillatory network that is a good example of 2-D cellular nonlinear network which is designed to generate spatial–temporal dynamical behavior. A cell in the network is a oscillator, and its model is originally inspired from ACE16k CNN-UM chip [17] and proposed by Yalcin in 2008 [22]. The network is given by

$$\dot{x}_{i,j} = \alpha x_{i,j} + \beta y_{i,j} + g(x_{i,j}) + w \Im_{i,j},$$
$$\dot{y}_{i,j} = \gamma x_{i,j} + \epsilon y_{i,j}, \tag{2.17}$$

where $x_{i,j}$ and $y_{i,j}$ are the state variables of the second-order cell $C_{i,j}$, the nonlinearity $g(x) = \mu(|x + \lambda| - |x - \lambda| - 2x)$, and the synaptic law is given by

$$\Im_{i,j} = \sum_{S_{i,j}(r)} A_{k-i,l-j} x_{k,l}, \tag{2.18}$$

where $S_{i,j}(r = 1)$ is given in Eq. (2.13). Each cell is corresponding to a relaxation oscillator, and its oscillating state ($\alpha = 4.2, \beta = 3, \gamma = 1, \epsilon = 0.5, \mu = -10, \lambda = 1$ and $w = 0$) in time domain is plotted in Fig. 2.5.

Locally coupled oscillatory network (2.17) is programable to generate autowave when the feedback template

$$A = \begin{bmatrix} 0 & 1 & 0 \\ 1 & 0 & 1 \\ 0 & 1 & 0 \end{bmatrix}, \tag{2.19}$$

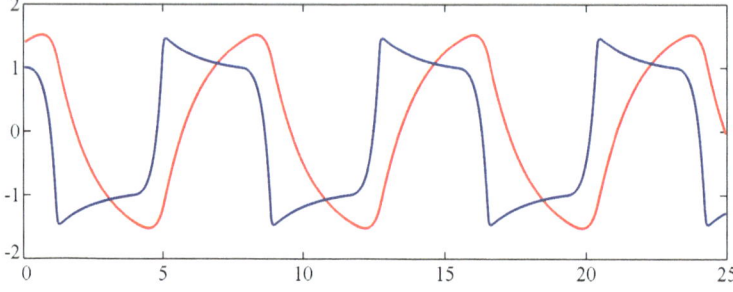

Fig. 2.5 $x(t)$ and $y(t)$ signals generated by an uncoupled cell of the network (2.17) with $\alpha = 4.2$, $\beta = 3, \gamma = 1, \epsilon = 0.5, \mu = -10, \lambda = 1$ and $w = 0$

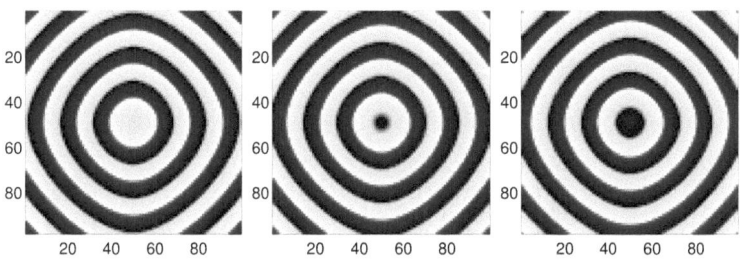

Fig. 2.6 Autowave propagation on 90×90 CNN (2.17) by the view of x states **a** at $t = 39$, **b** at $t = 40$, and **c** at $t = 41$

$w = 0.3$, and the other parameters are kept the same. In fact, the wave in spatial domain is a result of small phase differences between neighbor cells. In Fig. 2.6a–c, the x states of the network (2.17) are depicted in three separate images at times $t = 39, t = 40$, and $t = 41$, respectively. Here, the numerical solution of the network has been computed by forward Euler integration method with iteration time step $h = 0.1$. The waveform in the figure is called autowave due to its self-generating ability. Once an autowave is generated on such a network, it does not require to be fed by any input signal in order to generate successive wave fronts. And the waves do not reflect at the boundaries.

Locally coupled oscillatory network was used to solve 2D robot pathfinding problem using the wave fronts generated by the network in [23]. The arena of the robot is modeled as the medium of the waves on the network. The waves are employed to cover the whole medium with their dynamics, by starting from an initial point. The proposed path planning algorithm in [23] is achieved by observing the motion of the wave front of the waves. Kilic and Yalcin [24] improved this algorithm to perform in a 3D environment. In Sect. 4.2, an implementation of locally coupled oscillatory network model on Field-Programmable Gate Array (FPGA) which is a semiconductor device will be presented. Then, this implementation will be exploited to yield solution for motion planning problem in Sect. 4.2.2.

2.4 Modeling Wilson–Cowan Neuron Population with CNNs

2.4.1 One-layer CNN Model for Wilson–Cowan Neuron Population

Here, Wilson–Cowan neuron population model is rearranged such as a network of locally coupled inhibitory and excitatory cells. A cell in network might be a member of the set of inhibitory C^I or excitatory C^E neuron population. Randomness can be introduced to a locally coupled CNN topology by randomly locating excitatory and

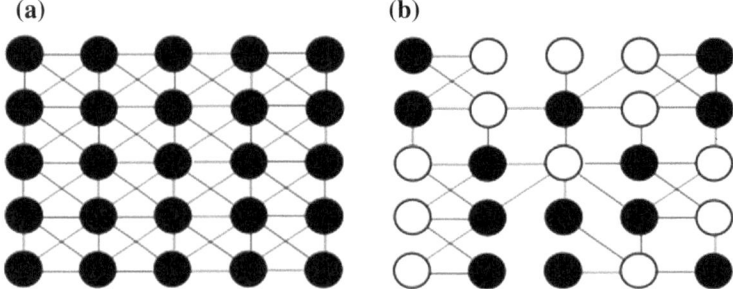

(a)   **(b)**

Fig. 2.7 2-D cellular nonlinear network with **a** same type of cells, **b** two types of cells which are randomly located in the array

inhibitory neurons. In 2-D CNN, a cell $C_{i,j}$ which is located in cartesian coordinate (i, j) might be a member of C^I or C^E. Figure 2.7 shows locally coupled and randomly located two types of cells in a CNN. Hence, each cell in the sets C^I and C^E is randomly located in 2-D array. One should note that when the cells are located in the cartesian coordinate, the architecture of the network together with connection configuration is fixed. A new network is required, if a different architecture will be tested to search a better result.

A cell in C^I is defined by the following state equation:

$$\beta_{i,j}^I \frac{dy_{i,j}(t)}{dt} = -y_{i,j}(t) + \mu_I + \mathfrak{I}_{i,j}^I, \tag{2.20}$$

with its the synaptic law

$$\mathfrak{I}_{i,j}^I = K_i^I \theta^I \left(\sum_{S_{i,j}(r)} \left\{ A_{k-i,l-j} x_{k,l} \right\} + g_{inp} U_{i,j} \right). \tag{2.21}$$

Connection inside the population is not allowed in Wilson–Cowan neuron population. Therefore, local coupling is weighted only the state values of excitatory cell in the sphere of influence $S_{i,j}(r)$ which is defined in Eq. (2.14). If the cell in the coordinate (i, j) is an excitatory neuron, its dynamic is given by

$$\beta_{i,j}^E \frac{dx_{i,j}(t)}{dt} = -x_{i,j}(t) + \mu_E + \mathfrak{I}_{i,j}^E, \tag{2.22}$$

where the synaptic law of the excitatory cell is

$$\mathfrak{I}_{i,j}^E = K_i^E \theta^E \left(-\sum_{S_{i,j}(r)} \left\{ A_{k-i,l-j} y_{k,l} \right\} + g_{inp}^E U_{i,j} \right). \tag{2.23}$$

Connections between excitatory cells are not allowed like the other population. Connection weight in the sphere of influence is in fact a space-invariant parameter which is determined by A. However, randomness due to the location of subpopulations creates a randomly connected topology. Figure 2.7b shows local connections (straight line) between cells that belong to different populations. The introduced one-layer CNN model will be tested as a feature extractor to achieve better performance for odor classification problem in Sect. 3.7.

In the following section, two-layer CNN model is introduced to keep space-invariant connection weight property of CNNs.

2.4.2 Two-layer CNN Model for Wilson–Cowan Neuron Population

For two-layer CNN, a layer of size $N \times M$ is built on a regular grid and duplicated with different cells. Consequently, each layer has an identical cell model in it and one can use the same and also different cell models in the layers. In order to realize Wilson–Cowan neuron population model on CNN architecture, two-layer cellular neural network is created by one layer of inhibitory cells and one layer of excitatory cells as shown in Fig. 2.8.

The layer which consists of excitatory cells is governed by the following equation:

$$\beta_{i,j}^E \frac{dx_{i,j}(t)}{dt} = -x_{i,j}(t) + \mu_E + \mathfrak{I}_{i,j}^E, \qquad (2.24)$$

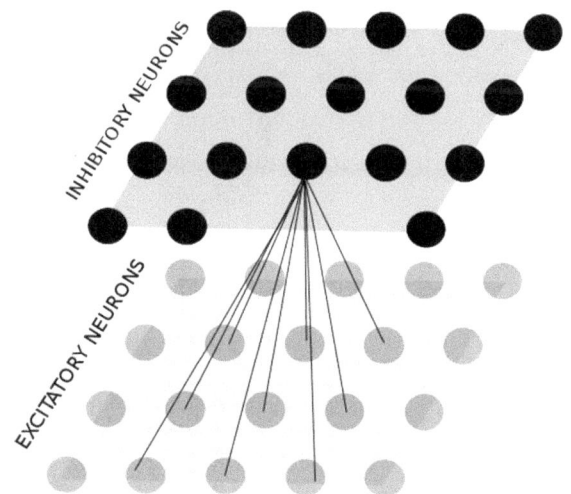

Fig. 2.8 2-layer CNN. Mathematical models of each cell are identical in the layer

where the synaptic law of the excitatory cell is given by

$$\mathfrak{I}_{i,j}^E = K_i^E \theta^E \Big(- \sum_{S_{i,j}(r)} \Big\{ A_{k-i,l-j}^E x_{k,l} + \hat{A}_{k-i,l-j}^E y_{k,l} \Big\} + g_{inp}^E U_{i,j} \Big). \qquad (2.25)$$

Dynamic of the second layer is given by

$$\beta_{i,j}^I \frac{dy_{i,j}(t)}{dt} = -y_{i,j}(t) + \mu_I + \mathfrak{I}_{i,j}^I, \qquad (2.26)$$

with its the synaptic law

$$\mathfrak{I}_{i,j}^I = K_i^I \theta^I \Big(\sum_{S_{i,j}(r)} \Big\{ \hat{A}_{k-i,l-j}^I x_{k,l} + A_{k-i,l-j}^I y_{k,l} \Big\} + g_{inp} U_{i,j} \Big). \qquad (2.27)$$

A connection inside the population is not allowed, and connection template for inhibitor layer (A^I) and excitatory layer (A^E) is given by a zero matrix. Connections between inhibitory and excitatory neuron are given with templates $\hat{A}^I$ and $\hat{A}^E$.

The first layer consists of inhibitory neuron population and the second layer contains only excitatory neuron population, so that all the neurons in the same CNN layer have the same mathematical model. One should note that each neuron might have different input signals; furthermore, it is not necessary that each cell has an input. These neurons are randomly located in each layer and each neuron in a subpopulation is locally coupled to its neighbors in the other subpopulation. Neighborhood is defined by the sphere of influence $S_{i,j}(r)$. Therefore, network connection topology has a regular architecture. Characteristic of randomness of Wilson–Cowan neuron population is mimicked by randomly locating neurons (which has different inputs) in two-layer CNN which is in fact not sufficient to have the same performance as the population model.

The test of this model to use as a feature extractor will be reported in Sect. 3.7.

2.5 Small-World Cellular Neural Networks (SWCNNs)

The small-world architecture model which is introduced by Watts and Strogatz [25] is a network consisting of many local links and fewer long-range shortcuts. Recent studies indicate that associative memory networks with small-world architectures and randomly connected networks have the same retrieval performance [26]. Therefore, SWCNN is a good candidate to model biological neural network which has randomly connected topology such as Wilson–Cowan neuron population.

Small-world cellular neural network is introduced by Tsurata et al. [27] in 2003. In fact, SWCNN is constructed by adding some random coupling between the cell which is not in its sphere of influence.

Dynamic of SWCNN is defined by Eq. (2.12) with the sphere of influence (2.13). The synaptic law of the corresponding cell of SWCNN is given by

$$\mathfrak{J}_{i,j} = \sum_{S_{i,j}(r)} \{A_{k-i,l-j}y_{k,l} + B_{k-i,l-j}U_{k,l}\} + w_{i,j;m,n}M(i,j;m,n)y_{m,n}, \quad (2.28)$$

where $n \in \{1, 2, ..., N\}, m \in \{1, 2, ..., M\}, M(i, j; m, n)$ is the small-world map that is randomly created with the probability p. Hence, the cells $C_{i,j}$ and $C_{m,n}$ are coupled with the probability p. The coupling weight between the randomly coupled cells is given by $w_{i,j;m,n}$. One should note that a regular CNN is obtained when $p = 0$.

Wilson–Cowan neuron population is modeled with 2-layer SWCNN by extending the network which is given in Sect. 2.8 [28]. Layer of excitatory neurons in 2-layer SWCNN is given by Eq. (2.24), and the synaptic law of a cell in this layer is

$$\mathfrak{J}_{i,j}^E = K_i^E \theta^E \Big(- \sum_{S_{i,j}(r)} \Big\{ A_{k-i,l-j}^E x_{k,l} + \hat{A}_{k-i,l-j}^E y_{k,l} \Big\}$$
$$+ g_{inp}^E U_{i,j} + w_{i,j;m,n}^E M^E(i,j;m,n)y_{m,n} \Big). \quad (2.29)$$

Layer of inhibitory neurons is defined by Eq. (2.26), and the synaptic law is

$$\mathfrak{J}_{i,j}^I = K_i^I \theta^I \Big(\sum_{S_{i,j}(r)} \Big\{ \hat{A}_{k-i,l-j}^I x_{k,l} + A_{k-i,l-j}^I y_{k,l} \Big\}$$
$$+ g_{inp} U_{i,j} + w_{i,j;m,n}^I M^I(i,j;m,n)x_{m,n} \Big). \quad (2.30)$$

Figures 2.8 and 2.9 show an inhibitory neuron $C_{i,j}$ in the same location and its connection between the cells in the other layer.

Fig. 2.9 2-layer small-world CNN

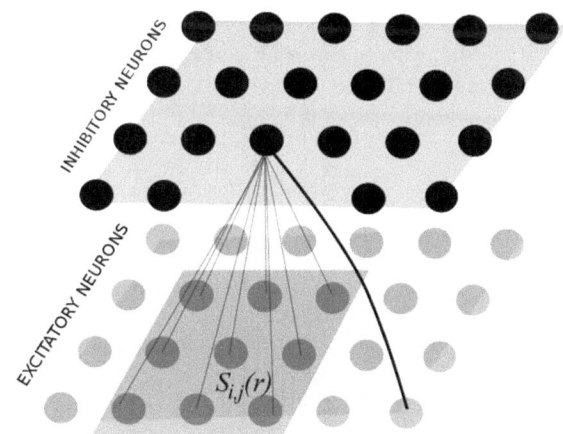

Random connections which are defined by the small-world maps $M^I(i, j; m, n)$ and $M^E(i, j; m, n)$ in the synaptic laws are unilateral. Taking $M^E(i, j; m, n) = M^I(i, j; m, n)$, bilateral random connection is obtained.

Multi-type processor networks can be obtained with multilayer architectures where each layer contains single type of processors. However, the topology of the layers is fixed after realization. In the next section, we offer to locate two types of dynamics on the same layer and propose a formulation that enables us to change the locations of the populations randomly even after the network is implemented.

The presented small-world CNN model will be tested to find out the effect of small-world phenomenon on odor classification problem in Sect. 3.8.

2.6 Reconfigurable Cellular Neural Network

In Reconfigurable Cellular Neural Network (RCNN), a cell has a set of dynamical systems with the same state variables and cells are able to program to be governed by one of the dynamic systems in the set. Therefore, RCNN is thought of as a cellular nonlinear network. RCNN was first introduced by Ayhan and Yalcin [29] to handle bioinspired sensory processing task of odor classification. The work of Ayhan and Yalcin [29] is given in the next section which is in fact a special case of the generalized model of RCNN given in this section.

A cell in a two-dimensional array of RCNN is here given by

$$
\begin{aligned}
\dot{x}_{i,j} &= g_{\text{ID}}(x_{i,j}, z, \mathfrak{I}_{i,j}, U_{i,j}) \\
v_{i,j} &= f_{\text{ID}}(x_{i,j}),
\end{aligned}
\tag{2.31}
$$

and its sphere of influence which defines a local coupling in radius r is given in Eq. (2.13). ID is identity variable for cells and dynamic of a cell is defined by ID where ID $\in \{0, 1, ..., L\}$. Figure 2.10 shows a 5×5 R CNN which has four different cell identities ($L = 4$). Each identity (ID) is indicated by a gray level in the figure.

One should note that each cell can be configured to have all possible identities. Therefore, a cell in RCNN will be denoted by $C_{i,j}(\text{ID})$.

The synaptic law of the corresponding cell $C_{i,j}(\text{ID})$ is given by

$$
\mathfrak{I}_{i,j} = \sum_{S_{i,j}(r)} \{\theta_{\text{ID}}(\hat{A}_{k-i,l-j})y_{k,l} + B_{k-i,l-j}U_{k,l}\}.
\tag{2.32}
$$

Local coupling weight might also be changed by cell identity. For example, in Fig. 2.10, cells with the same cell ID are not connected between each other, and the only connection between the cell ID 0 and the other cell ID is allowed.

The types of the neurons in a network are defined with the ID matrix H and its dimension is equal to the dimension of network. $h_{i,j}$ are identity number and $h_{i,j} \in \{0, 1, ..., L\}$. The network in Fig. 2.10a is configured with ID matrix

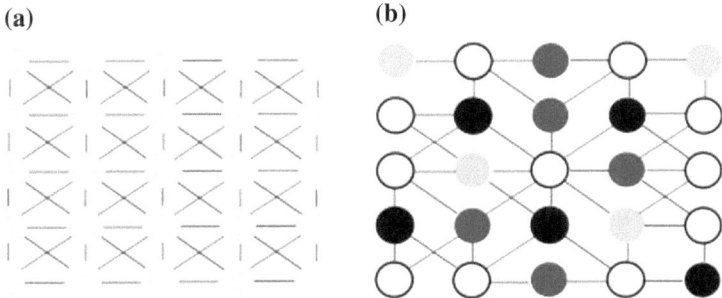

Fig. 2.10 A 5×5 **a** Reconfigurable Cellular Neural Network (RCNN) before the reconfiguration **b** and after the reconfiguration the with four different cell types ($L = 4$). Gray level indicates cell's identity and the identity (ID) H is given in Eq. (2.33)

$$H = \begin{bmatrix} 1 & 0 & 2 & 0 & 1 \\ 0 & 3 & 2 & 3 & 0 \\ 0 & 1 & 0 & 2 & 0 \\ 3 & 2 & 3 & 1 & 0 \\ 0 & 0 & 2 & 0 & 3 \end{bmatrix}, \tag{2.33}$$

and the obtained network is shown in Fig. 2.10b. The network is reconfigured by different H matrices.

RCNN allows collaboration of L number of subpopulations which is in fact the key concept in processing and decision mechanisms of biological networks. The connection topology of RCNNs which is still defined by the sphere of influence $S_{i,j}(r)$ and deterministic becomes random by randomly generated ID matrix H. While RCNN architecture allows to configure different subpopulations on itself, responses of RCNN for randomly generated H can be used to obtain the best result for a given task. This is in fact allows us to have a kind random search algorithm which generally provides a relatively good solution.

In the next section, RCNN will be used to model Wilson–Cowan neuron population network by randomly generated ID matrix. Hence, two different types of neurons are randomly located in a single locally coupled network layer.

2.6.1 RCNN Model of Wilson–Cowan Neuron Population

In order to realize neural mass models, cellular neural networks should imitate the dynamics of excitatory and inhibitory neurons. The types of the neurons in a network are defined with the ID matrix H of size $M \times N$. $h_{i,j}$'s are either 1 or 0 corresponding to excitatory and inhibitory neurons, respectively. Characterization of cells on a regular CNN grid is shown in Fig. 2.11.

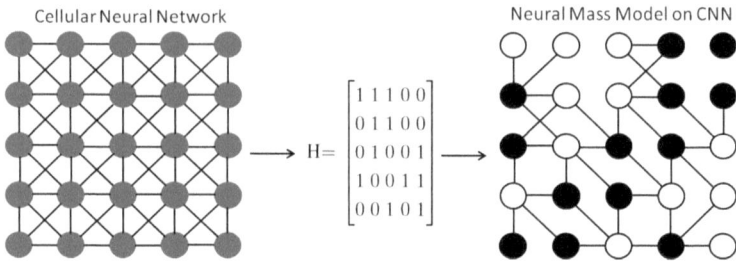

Fig. 2.11 In A 2-D CNN layer with all identical processor units, the cells (gray) have the same dynamics. ID matrix H is used to characterize them. After applying H to the grid, the cells behave as either inhibitory (black) of excitatory (white)

Difference between excitation and inhibition is only a matter of sign in the activation level—synaptic law. The effect of signature is reflected to synaptic law by ID matrix H:

$$\Im_{i,j} = g\big(H, A, x_{i,j}, U_{i,j}\big) \tag{2.34}$$

for Wilson–Cowan population model, and synaptic law is given by

$$\Im_{i,j} = K\theta\bigg((h_{i,j} - h_{k,l}) \sum_{l=-r}^{r} \sum_{k=-r}^{r} a_{k,l}.x_{i+k,j+l}(t) + g_{inp}U_{i,j}\bigg), \tag{2.35}$$

where $U_{i,j}$ is the sensory input, g_{inp} is the input weight, and $a_{k,l}$ is drawn from state-feedback template A. In this case, the *sphere of influence* $S_{i,j}(r)$ of radius r is defined as in Eq. (2.10). Dynamic of a cell in RCNN model is given by

$$\beta_{i,j} \frac{dx_{i,j}(t)}{dt} = -x_{i,j}(t) + \mu + \Im_{i,j}. \tag{2.36}$$

If $h_{i,j}$ is 1 for instance, Eq. (2.36) will be equal to dynamic of excitatory neurons and it will only connect to inhibitory neurons in its sphere of influence $S_{i,j}(r)$. The CNN layer can still exhibit different behaviors with same type of processors. In other words, with an appropriate modification in synaptic law, the traditional CNN layer of identical processors runs like a many-type processor network. Moreover, the locations of excitatory and inhibitory neurons are determined by ID matrix and it can be used for reconfiguring the network. Two key properties of biological networks are (i) different dynamics on the same network and (ii) randomness in connections, added to CNN architecture with RCNN model. From computational point of view, the locations of inhibitory and excitatory neurons are changed with ID matrix in run time in order to obtain the best result.

In Sect. 3.9, RCNN model will be used as a feature extraction to accelerate the odor classification problem and test results will be presented. Similar feature extraction on RCNN model will be studied for EEG signal classification problem in Sect. 3.9.1. The role of ID matrix which describes network topology will be studied in Sect. 3.10. Furthermore, an implementation of RCNN model on Field-Programmable Gate Array (FPGA) will be presented in Sect. 4.4.

References

1. F. Rosenblatt, *Principles of Neurodynamics: Perceptrons and the Theory of Brain Mechanisms* (Spartan Books, 1962)
2. J.J. Hopfield, Neural networks and physical systems with emergent collective computational properties. Proc. Nat. Acad. Sci. **79**, 2554–2558 (1982)
3. L.O. Chua, L. Yang, Cellular neural networks: theory and applications. IEEE Trans. Circuits Syst. I **35**(10), 1257–1290 (1988)
4. H.R. Wilson, J.D. Cowan, A mathematical theory of the functional dynamics of cortical and thalamic nervous tissue. Kybernetik **13**, 55–80 (1973)
5. H. Hartline, F. Ratlif, Spatial summation of inhibitory in influences in the eye of limulus, and the mutual interaction of receptor units. J. Gen. Physiol. **41**, 1049–1066 (1958)
6. F.H.L. Da Silva, A. Hoeks, H. Smits, L.H. Zetterberg, Model of brain rhythmic activity. Biol. Cybern. **15**, 27–37 (1974)
7. W.J. Freeman, *Mass Action in the Nervous System* (Academic Press, New York, 1975)
8. M.K. Muezzinoglu, A. Vergara, R. Huerta, T. Nowotny, N. Rulkov, H.D.I. Abarbanel, A.I. Selverston, M.I. Rabinovich, Artifcial olfactory brain for mixture identifcation, in *Neural Information Processing Systems* (2008), pp. 1121–1128
9. W. Freeman, A neurobiological theory of meaning in perception. Part I: information and meaning in nonconvergent and nonlocal brain dynamics. Int. J. Bifurc. Chaos. **13**(9), 2493–2511 (2003)
10. D. Watts, S. Strogatz, Collective dynamics of 'small-world' networks. Nature **393**(6684), 440–442 (1998)
11. S.I. Amari, Dynamics of pattern formation in lateral-inhibition type neural fields. Biol. Cybern. **27**(2), 77–87 (1977)
12. L.B. Vosshall, A.M. Wong, R. Axel, An olfactory sensory map in the fly brain. Cell **102**(2), 147–159 (2000)
13. L.O. Chua, *CNN: A Paradigm for Complexity* (World Scientific, Singapore, 1998)
14. M.E. Yalçın, J.A.K. Suykens, J. Vandewalle, *Cellular Neural Networks, Multi-scroll Chaos and Synchronization* (World Scientific, 2005)
15. S. Espejo, C. Carmona, R. Dominguez-Castro, A. Rodriguez-Vazquez, CNN universal chip in CMOS technology. Int. J. Circuit Theory Appl. **24**, 93–111 (1996)
16. T. Roska, L. Chua, The CNN universal machine—an analogic array computer. IEEE Trans. Circuits Syst. II Analog. Digit. Signal Process. **40**(3), 163–173 (1993)
17. A. Rodriguez-Vazquez, G. Linan-Cembrano, L. Carranza, E. Roca-Moreno, R. Carmona-Galan, F. Jimenez-Garrido, R. Dominguez-Castro, S. Meana, ACE16k The third generation of mixed-signal SIMD-CNN ACE chips toward VSoCs. IEEE Trans. Circuits Syst. I Regul. Pap. **51**(5), 851–863 (2004)
18. Cellular Sensory Wave Computers Laboratory: Cellular Wave Computing Library. Computer and Automation Research Institute-Hungarian Academy of Sciences (2007)
19. P. Foldesy, L. Kek, A. Zarandy, T. Roska, G. Bartfai, Fault-tolerant design of analogic CNN templates and algorithms-part I: the binary output case. IEEE Trans. Circuits Syst. I Fundam. Theory Appl. **46**(2), 312–322 (1999)
20. S. Xavier-de-Souza, M.E. Yalcin, J.A.K. Suykens, J. Vandewalle, Toward CNN chip-specific robustness. IEEE Trans. Circuits Syst. I Regul. Pap. **51**(5), 892–902 (2004)

21. E. Kose, M.E. Yalcin, A new architecture for emulating CNN with template learning on FPGA, in *The 16th International Workshop on Cellular Nanoscale Networks and their Applications (CNNA 2018)* (2018), pp. 1–4
22. M.E. Yalcin, A simple programmable autowave generator network for wave computing applications. IEEE Trans. Circuits Syst. II Express Briefs. **55**(11), 1173–1177 (2008)
23. R. Yeniceri, M.E. Yalcin, Path planning on cellular nonlinear network using active wave computing technique, in *Proceedings of SPIE, Bio-engineered and Bioinspired Systems IV*, vol. 7365 (2009)
24. V. Kilic, M.E. Yalcin, An active wave computing based path finding approach for 3-D environment, in *Proceedings of the IEEE International Symposium of Circuits and Systems (ISCAS 11)* (2011), pp. 2165–2168
25. D. Watts, S. Strogatz, Collective dynamics of small-world networks. Nature **393**, 440–442 (1998)
26. J.W. Bohland, A.A. Minai, Efficient associative memory using small-world architecture. Neurocomputing **3840**, 489–496 (2001)
27. K.Y.N. Tsurata, Y. Zonghuang, A. Ushida, Small-world cellular neural networks for image processing applications, in *Proceedings of European Conference on Circuit Theory and Design* (2003), pp. 225–228
28. T. Ayhan, M.E. Yalcin, An application of small-world cellular neural networks on odor classification. Int. J. Bifurc. Chaos. **22**(1), 1–12 (2012)
29. T. Ayhan, M.E. Yalcin, Randomly reconfigurable cellular neural network, in *Proceedings of the 20th European Conference on Circuit Theory and Design (ECCTD11)* (2011), pp. 625–628

Chapter 3
Artificial Olfaction System

3.1 Introduction

Like audition and vision, olfaction provides most animals information about their surroundings. This information is gathered from odors which are a sensation caused by odorants around. Many objectives may be introduced in order to solidify the necessity of odor classification by electronic terms, but the main motivation behind this need is the fact that both human and animals use the knowledge of odor from feeding and sexual behavior to identifying objects that are potentially dangerous such as fire, gas, and poisonous food. Odor information is widely used for various purposes. For example, for discrimination of two objects with similar appearances like water and alcohol, smelling is safer than tasting. Also for determining the quality of an object, usually food, smelling is preferred to taste. Since olfaction system of dogs is very advanced, dogs are used for target tracking and search of nonmetal objects like heroin and living organisms. Smelling is also a part of medical in the diagnosis of certain respiratory and digestive system illnesses such as diabetes.

In order to provide olfactory information and process odors artificially, machine olfaction gave rise to developments in biological modeling, sensor technology, and bioinspired technologies. A bioinspired application on olfaction is an electronic nose. In the past decade, electronic nose instrumentation has generated much interest internationally for its potential to solve a wide variety of problems in fragrance and cosmetics production, food and beverages manufacturing, chemical engineering environmental monitoring, and more recently, medical diagnostics and bioprocess [1]. An electronic nose is a machine that is offered to detect the odorants around and process them to classify or give a decision.

Sensing is realized by various methods and advances in sensing depend on the developments of sensor technology. Sensor technology and instrumentation for electronic noses are commercially available but usually in desktop form, not suitable for mobile uses because of size, sensing abilities or power consumption. Since sensing part is a built-in system, pattern analysis and signal processing techniques can be applied by advanced processors, so algorithmic cost of these methods is not an issue in odor processing. However, odor sensory data is relatively new in signal processing era and olfactory processing deviates from auditory and visual signal processing because of the nature of the signal. Reason of principle difference lies on the sensory

M. E. Yalçın et al., *Reconfigurable Cellular Neural Networks and Their Applications*, SpringerBriefs in Nonlinear Circuits, https://doi.org/10.1007/978-3-030-17840-6_3

level. Sensors are short-lived, produced with high mismatch or are too slow to set on a stable response. Therefore, odor processing also aims to overcome these problems caused by odor sensors.

In this chapter, the data collected with commercially available metal oxide gas sensors of Figaro Inc. are used. Then the sensory data is processed with cellular neural network-based artificial antennal lobes on the transient regime in order to accelerate the decision interval and decrease the time required for classification. Not only the processing unit but also some sensory characters may have a positive effect on decision time so sensor temperature is also altered and its impact on classification performance is discussed.

3.2 Biological Olfaction Systems

An odor is a perception that arises when an odorant which is a compound generates stimuli in the receptory part of the olfaction system of an animal [1]. Like vision and audition, olfaction is vital for analyzing the surroundings odor signals of food, environment, or other organisms that carry concise information about feeding, sexual behavior, and potential danger which are vital.

Odorants are usually organic compounds which are volatile and hydrophobic as well as some inorganic compounds such as ammonia and hydrogen sulfide. It is important to note that, although the concentration of the odorant is very low in the air (in the orders of parts-per-billion or even parts-per-trillion) it can still be detected [2]. Therefore, not the concentration of the odorant but the molecular shape of it makes the odor recognizable. A trained human olfaction system is capable of distinguishing more than 10,000 odors [3]. However, not all the objects have an odor, but some compounds which have a molecular weight less than 300 Dalton and properties such as water solubility, sufficient water pressure, low polarity some ability to dissolve in fat (lipophilicity), and surface activity evoke sensory process [3]. Discovery of Olfactory Receptor (OR) genes by Buck and Axel [4] accelerated the researches on both physiological and biochemical behavior of the olfaction system and a great advance has been accomplished which leads the researches toward an artificial olfaction system. Odors can be labeled with the sensations that they raise for human, such as sweet or acidic.

In vertebrates or in insects, stimuli to odorants are created on a similar path which starts at the Olfactory Receptors (ORs). Odorants are caught by ORs which are located on the olfactory epithelium in the nasal cavity in vertebrates or by the antenna in insects. A preprocessing unit called antennal lobe (for insects) or olfactory bulb (for human) follows the sensory part and leads the stimuli to the higher levels of the nervous system for any type of decision such as classification.

3.2.1 Olfaction System of Insects

Having fewer neurons and receptors, insects are better models to imitate olfactory systems than more complex mammals. Studies on Drosophila give much information about the olfactory system in insects. Olfactory system in insects is composed of three stages as shown in Fig. 3.1. A sensory encoding of the stimuli by a vast diversity of receptors is captured by the projection neurons and local neurons which generate a spatiotemporal pattern. This pattern is read out by the mushroom body in the form of static images (i.e., snapshots). This final layer is considered to be the generic classifier of the insect brain.

The first stage of olfaction is the antenna where odor information is gathered by the olfactory receptors. Although olfactory neurons which express the same kind of odor may include more than one neuron, they are all connected to the same glomeruli [5]. Glomeruli provides a connection between olfactory sensory neurons and local neurons which are primarily inhibitory and projection neurons [6]. Odor discrimination is processed in higher layers of the olfaction system where information is projected from antennal lobe by projection neurons. Therefore, the effect of odorants on antenna neurons (primary neurons) is somehow moved to higher order neurons for discrimination [5]. One of the key points of olfaction in insects is how the receptors are connected to classification units called mushroom bodies. This projection is done in the antennal lobe, the second stage of the olfactory system.

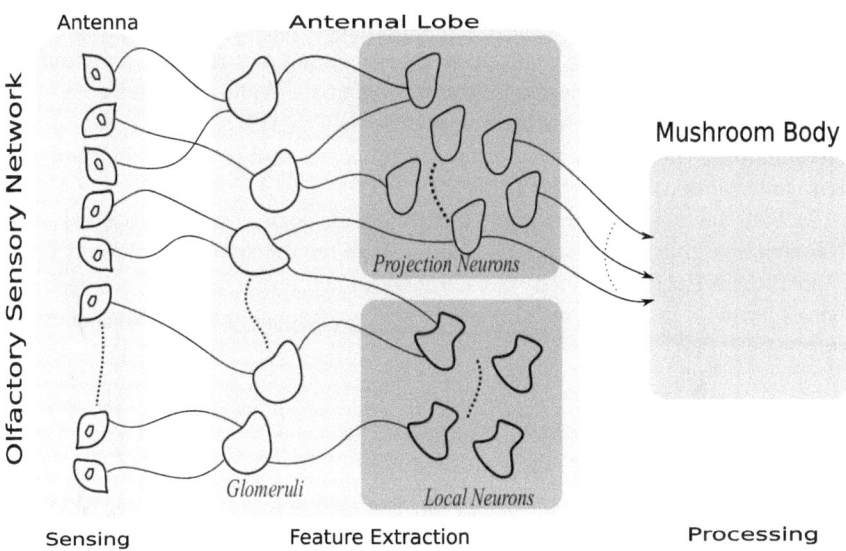

Fig. 3.1 A block diagram interpretation of the generic insect olfactory system and functionality of the corresponding subblock of the system

3.2.2   Antennal Lobe

Antennal lobe is in the microbrain of the insect. Its behavior and function have similarities with the olfactory bulb in vertebrates. Antennal lobe is an intermediate region between the data collecting olfactory sensory neurons on the antenna and higher brain centers, such as the mushroom body. Antennal lobe acts as a preprocessing unit, which converts the sensory neuron data into a spatiotemporal code. This conversion happens at the intersection point of the olfactory receptor neurons, local neurons of the antennal lobe, and the projection neurons that are responsible for projecting the data to higher brain centers. This neuron synapsing region is called glomeruli and the antennal lobe is composed of many of them. Generally, each projection neuron in Drosophila receives information from a single glomerulus. Coding properties and learning (plasticity) capacities of the projection neurons are timely research topics. Authors refer to [7] for more information on this issue.

3.2.3   Mushroom Body

Getting their name from their calyx-like structure, mushroom bodies (or corpora pedunculata) are located in the brain of insects and other arthropods. These structures, where neurons and synapses form sense networks, were first identified in 1850 by the French biologist Flix Dujardin [8]. About their function, mushroom bodies are associated with learning and memory, especially with odor-related memory. For larger insects, mushrooms bodies have other learning and memory functions too; therefore, they resemble cerebral cortex of mammals. Although mushroom bodies are not as big as the vertebrate brain structures, they are still able to process complex learning [8]. Therefore, genetic control of processes within the mushroom bodies arise interest among researchers.

Leaving the larger insects apart, smaller insects such as the fruit fly Drosophila melanogaster give the primary information about the function of mushroom bodies. Since the genetics of this fruit fly is extensively known, research on fruit fly mushroom bodies provides important insights on the genetic basis of their functioning.

3.2.4   Olfaction System of Mammals

The olfactory system of many mammals contains two distinct subsystems: main olfactory system, which detects volatile stimuli of odorants in the air, and an accessory olfactory system, which detects fluid-phase stimuli and especially evolved for sexual behavior. In this subsection, the main olfactory system will be investigated.

The olfaction system works in harmony with trigeminal receptors on the tongue to enjoy the flavor because human tongue can distinguish only among a few types of taste where nose can distinguish among more [9]. However, this secondary function of olfaction system will not be covered here. In air-breathing animals, the nose is covered by two types of epithelium partially: olfactory epithelium, where the ORs are located, and the respiratory epithelium. The ratio between them gives an insight into the olfactory sensitivity of the animal. In order to compare the olfactive sensitivity, some dogs have $170 \, cm^2$ epithelium with more densely located receptors, and they can distinguish up to 10,000 different odorants [9]. In humans, the olfactory epithelium covers a surface of about $3 \, cm^2$ and is composed of three main cellular types: the olfactory sensory neurons (also called olfactory receptors), the sustentacular or supporting cells, and the basal epithelial or stem cells [10].

Sensing unit of air-breathing mammals is placed in the nasal concha in the nasal cavity. Odorants dissolved or cached by the mucus lining in the nose and detected by ORs on the dendrites of the Olfactory Sensory Neurons (OSNs). Detection is a process of binding the odorant to odorant binding proteins [4].

The odor receptors are particularly interesting, as they are directly interacting with the stimuli generating odor molecules. While passing through the nose, odor molecules (or odorants) dissolve in the mucus which covers a large area in the nasal cavity. This way, the olfactory sensory neurons can detect the odorant on their dendrites. The receptor neuron reacts by creating an action potential when the receptor is bound with an odor molecule. Olfactory sensory activity in the sensory neurons of the insects and in the receptor neurons of vertebrates is measured by an electro-olfactogram [10]. Monitoring is done through calcium imaging or receptor neuron terminals in the olfactory bulb or within the olfactory bulb, for insects.

3.3 Short Overview of Artificial Olfaction System

Roughly, olfaction system is composed of three major parts which perform three duties:

- **Sensing**: Reacting to odorants around either with electrochemical signals.
- **Feature Extraction**: Changing spatial sensor response into spatiotemporal patterns.
- **Processing**: Evaluation of produced patterns: classification and decision.

Smelling begins at odor sensory neurons with the perception of volatile odorants and producing an electrochemical response. Like there exist different kinds of sensory neurons responding diversely to different environments; there are several metal oxide sensors types classified according to doped metal. Either in antenna or nose, odor sensory neurons employing the same kind of odor binding protein meet in the same glomeruli. As it is shown in Fig. 3.2, perception layer of an artificial nose is sensor array, and it is matched with odor sensory neurons together with glomeruli. The upper layer is named as feature extraction although a complete model for this section is not introduced yet its function on odor processing is to project suitable features to higher nervous system levels. The most important benefit of this layer is to fasten the

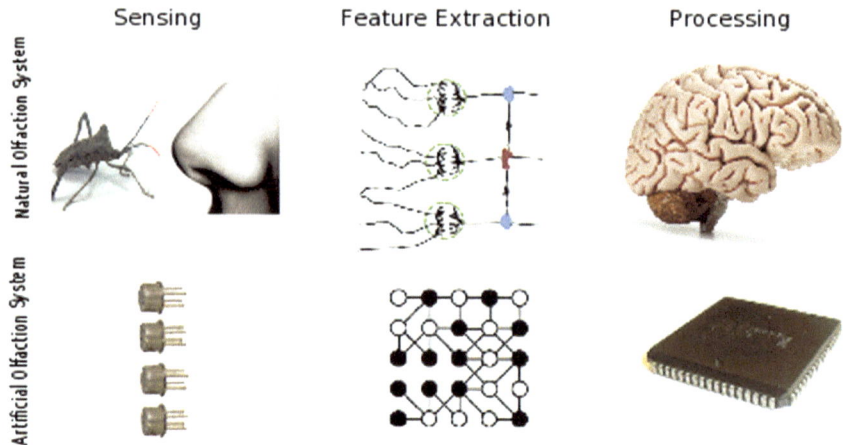

Fig. 3.2 Analogy between an artificial olfaction system and a natural one

decision by going beyond the limits of sensory neurons. Generation of spatiotemporal patterns from spatial sensor response is realized by CNNs in this book.

The processing unit shown in olfaction diagram can either be part of the olfaction system like olfactory cortex or any other part in brain. Therefore, that part does not benefit on olfaction system directly but uses the olfactory information. In this book, the processing unit is assumed to be a classifier in order to test the models.

3.4   Test for Proposed Models

Previously introduced CNN models in Chap. 2 will be tested here. The key point of the models is that they are designed to convert spatial sensor response into spatiotemporal patterns; this is usually done by antennal lobe in the insect's olfactory system. The goal of the antennal lobe models includes two branches: one is to achieve better performance in classification of odors according to the chosen problem. The other one is to reach a satisfactory high performance earlier in the transient.

The purpose of the tests is to check the validation of the models with its biological background. Note that the proposal is not to build a complete artificial olfaction system, but in order to display the act of artificial antennal lobe models clearly, a complete artificial olfaction system is drawn for some of the tests below, in order to examine the performance dependent on the process unit shown in Fig. 3.3. Support vector machine (SVM) [17] is used as a classifier with two different training algorithms. However, the efficiency of the artificial antennal lobe should be reported independent from the classification tool so the success of it in increasing and accelerating the separability of sensor data is tested. Principal Component Analysis (PCA) [11] and Mutual Information (MI) [12] are the two such methods and widely used for sensory data classification problems. The uses of SVM and PCA will be explained

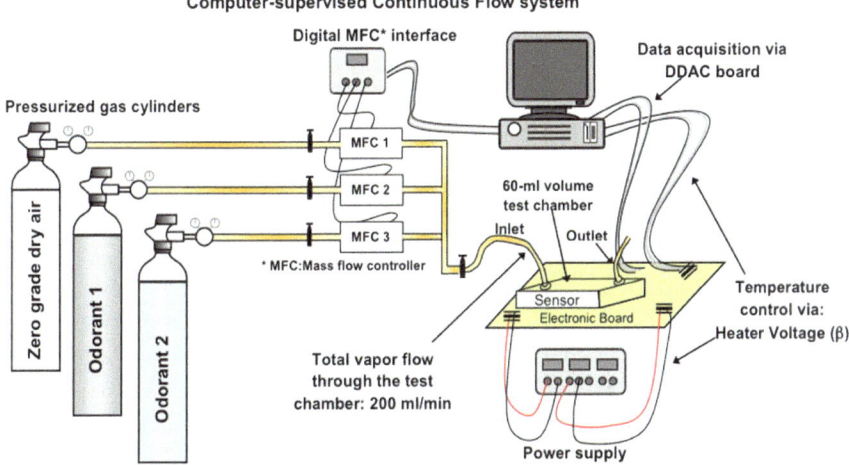

Fig. 3.3 Measurement setup in the laboratory of BioCircuits Institute, University of California San Diego (Reprinted from Vergara et al. [13], copyright 2012, with permission from Elsevier). The data sets which are used in this chapter are obtained from the real measurements

in Sects. 3.7.2 and 3.7.1, respectively. Input data are drawn from real measurements done in BioCircuits Institute, University of California San Diego. In this section, an overview of measurement setup is given, and the results of performance test with and without classification tool are reported.

3.5 Measurement Setup

Measurement setup [13] is given in Fig. 3.3. All the system is computer-supervised by LabVIEW. The odorants with desired concentration is streamed into the sensor chamber where 16 metal oxide gas sensors of Figaro Inc., namely, TGS2600, TGS2602, TGS2610, and TGS2620 [14] are placed. These sensors are commercially available and have many advantages over other types of odor sensors like QCM and SAW; they are easier to settle and use, and the mismatch is lower and they have a longer lifetime. Heater voltage of the sensors is controlled by LabVIEW® in order to adjust the internal temperature of the sensors. The odorants (gases) and dry air are stored in pressurized gas cylinders, and the amount of them streaming to the sensor chamber is controlled by Mass Flow Controllers (MFC) via digital MFC interface. Mass flow control is used to compose a mixture of desired ratios or adjust the concentration of target odor. The flow of vapor through the chamber is kept constant by MFC. The sensor activation is converted into a resistive signal by a simple electrical circuit on electronic board and signal is collected from the board. Slow time response is the disadvantage of the gas sensors. Therefore, in the artificial olfaction systems that use MOS sensors as sensing section, a feature extractor that uses the transient response of the sensor response is necessary.

3.6 Problem Definition

With the measurement setup given in the previous section, two problem sets are gathered. The first one, **Set A**, involves three classes: pure acetaldehyde, pure toluene, and mixture of those two. The concentration of the gas is adjusted by the dry air so a vapor of desired concentration is achieved. Vapor from one class is steamed into the test chamber under constant temperature (40 °C) for 100 s and the sensor data is collected with a sampling frequency of 100 Hz after injection. After injection, the chamber is cleaned with dry air flow. Twenty measurements are recorded for both pure acetaldehyde and pure toluene, 10 times for each of acetaldehyde–toluene mixtures which have the following compositions: 96−4%, 98−2%, 2−98%, 4−96%.

Examples of measurements, one of each class, is shown in Fig. 3.4. The data set is 80 times repeated records which offset is removed from each. Therefore, a data set of three classes, two pure odors and mixtures of them as a third class, is created.

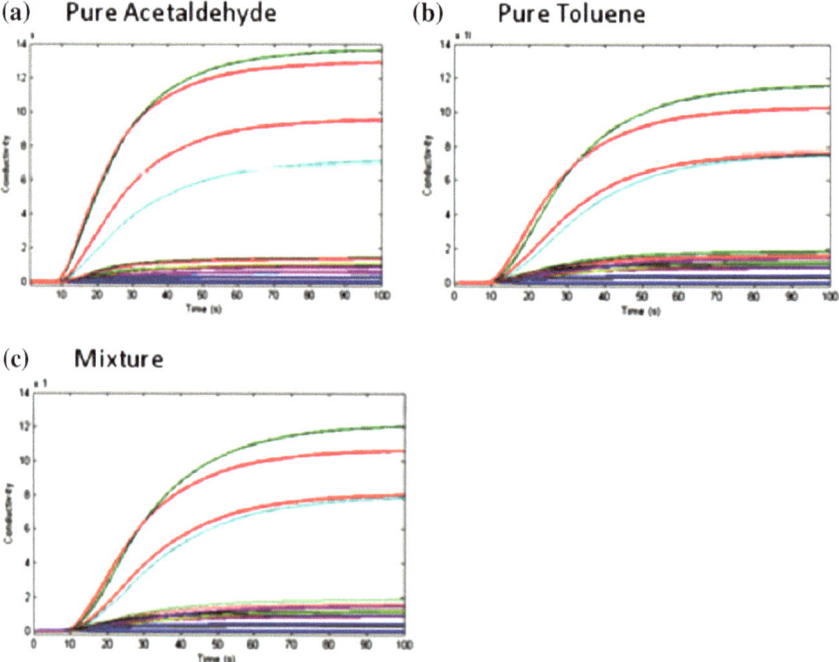

Fig. 3.4 Sensor response for **Set A**. An array of 16 sensors which consists of four of each TGS2600, TGS2602, TGS2610, and TGS2620 type metal oxide gas sensors. Sensors' responses for **a** pure acetaldehyde, **b** pure toluene, and **c** acetaldehyde–toluene mixtures

Set B contains five types of odor. A sensor array containing 16 metal oxide sensors (same measurement setup in **Set A**) is used to collect information of the following odors: acetaldehyde, acetone, butane, ethylene, and ethane. The sensors in the gas chamber are exposed to those gases one by one under constant flow and temperature for 100 s after the chamber is cleaned with dry air. Fifteen measurements for each odor are recorded.

Another data set, **Set C** is created for another three-class problem but is rather simple because three different odors are used. In this data set, the number of sensors in the array is decreased, and the performance of the sensor array is aimed to be increased by changing the sensor temperature. The effect of sensor temperature on performance is investigated in the next section.

3.7 Tests for Cellular Neural/Nonlinear Network-Based Models

Highly correlated variables do not increase the distinguishability of a data set while they carry more redundant information than uncorrelated variables. It is possible to foresee the separability of a data set by looking at statistical information such as a correlation between the variables.

CNN-based models are tested using **Set A** and with two classification tools, namely, Principal Component Analysis (PCA) and Support Vector Machines (SVM). Each measurement of data **Set A** is expanded with white noise with a variance of 0.5 and applied to CNN-based models: the models are driven by random noise for 50 s, and the measurement is applied at time zero. After 100 s, the system is continued to be driven by noise for 50 more seconds.

Data **Set A** is applied to one-layer CNN (described in Sect. 2.4.1) and two-layer CNN (described in Sect. 2.4.2), and the results of excitatory processors are collected as antennal lobe model output, only the activity of projection neurons are transferred to the higher layer and projection neurons are represented as excitatory cells in the models.

For one-layer CNN, a network of 10×15 is created with the connections explained in Sect. 2.4.1, and the half of the cells is labeled as excitatory and the other half is inhibitory. Therefore, 75 of the cells are excitatory and their output is recorded. In Fig. 3.5, the output of 75 excitatory neurons for pure acetaldehyde is given. The measurement is exposed from time 0 to 100.

In order to discuss the performance of CNN-based models in terms of connections, another model named uncoupled model is used for comparison. This uncoupled model involves 75 excitatory and 75 inhibitory cells which perform the same dynamics with the proposed two-layer CNN with $A^I = \hat{A}^I = 0$ and $A^E = \hat{A}^E = 0$ (defined in Sect. 2.4.2). The dynamics of uncoupled two-layer CNN is given by

$$\beta_{i,j}^I \frac{dy_{i,j}(t)}{dt} = -y_{i,j}(t) + \mu_I + K_i^I \theta^I \left(g_{inp} U_{i,j} \right) \tag{3.1}$$

Fig. 3.5 Output of one-layer
CNN for pure acetaldehyde

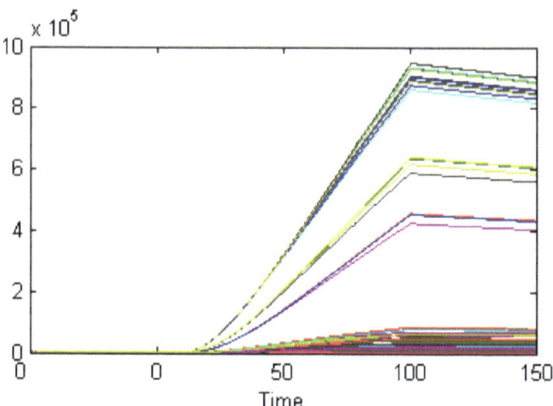

and

$$\beta_{i,j}^{E} \frac{dx_{i,j}(t)}{dt} = -x_{i,j}(t) + \mu_E + K_i^E \theta^E \left(g_{inp}^E U_{i,j} \right). \qquad (3.2)$$

3.7.1 Principal Component Analysis

Principal Component Analysis (PCA) is a simple method for extracting appropriate
information from complex and high order data sets such as sensor array responses.
Therefore, PCA is a widely used method for discriminating odors by sensor array
process. Some recent research on odor classification based on PCA using MOS
sensors is given in [15]. PCA was invented in 1901 by Karl Pearson [16]. This
data analysis tool reexpresses the data set with the most meaningful information by
calculation of the eigenvalue decomposition of a data covariance matrix. The new
data set is named with component scores and loadings and they have no dimension.

Mathematical definition of PCA is given in [11] as an orthogonal linear transfor-
mation that transforms the data to a new coordinate system such that the greatest
variance by any projection of the data comes to lie on the first coordinate (called
the first principal component), the second greatest variance on the second coordinate, and so on. First and second principal components for data **Set A** are shown in
Fig. 3.6. The x-axis is the time axis because the change of principal components in
time determines the increase in discrimination between classes in time.

As shown in Fig. 3.7, principal components belonging raw data (sensor array
output) for three classes do not show significant difference but both of the CNN-
based artificial antennal lobe model outputs are in the form that can be clustered in
this early time in the transient and the in the case for uncoupled model, principal
components of pure acetaldehyde can be allocated from other two classes already.

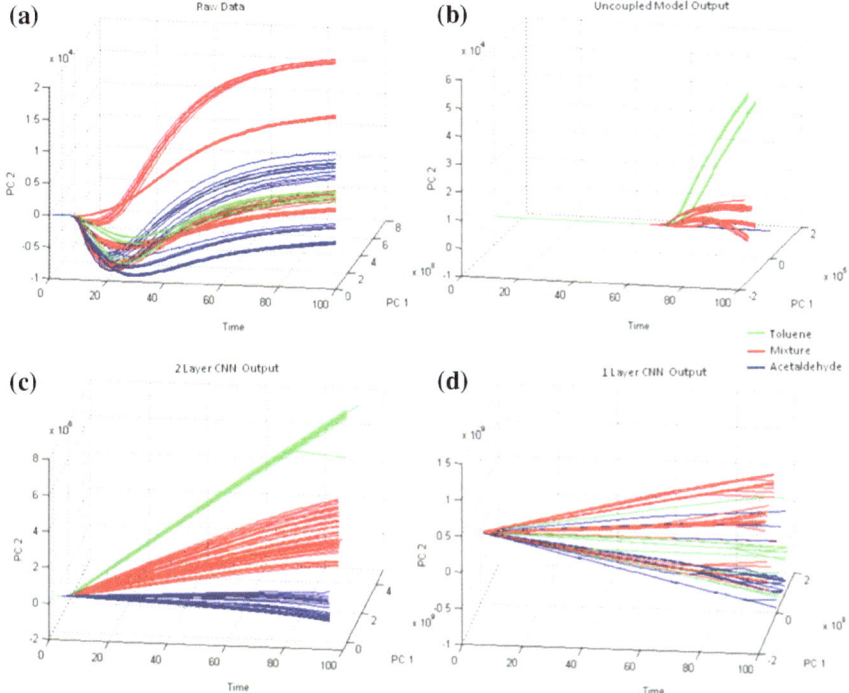

Fig. 3.6 Change of principal components in time **a** for raw data, **b** for uncoupled model, **c** for two-layer CNN-based artificial antennal lobe, and **d** for one-layer CNN-based artificial antennal lobe

The principal components for $T = 100$ s, which means the steady-state sensor response is given in Fig. 3.8. Classes can be discriminated from each other by clusters for both of the four graphs.

In order to give detail about the change of principal components in time, Fig. 3.9 is given. There, it can be clearly seen that classes have become separable by $T = 25$ s for two-layer CNN output.

3.7.2 Support Vector Machine

Support Vector Machine (SVM) pioneered by Cortes and Vapnik [17] is a supervised learning method used for classification or regression like multilayer perceptrons and radial-basis function networks. For this problem, it is used as a classification tool. Basically, support vectors assist to data points that are not linearly separable to be mapped in another space where they are separable. Therefore, an SVM model is a

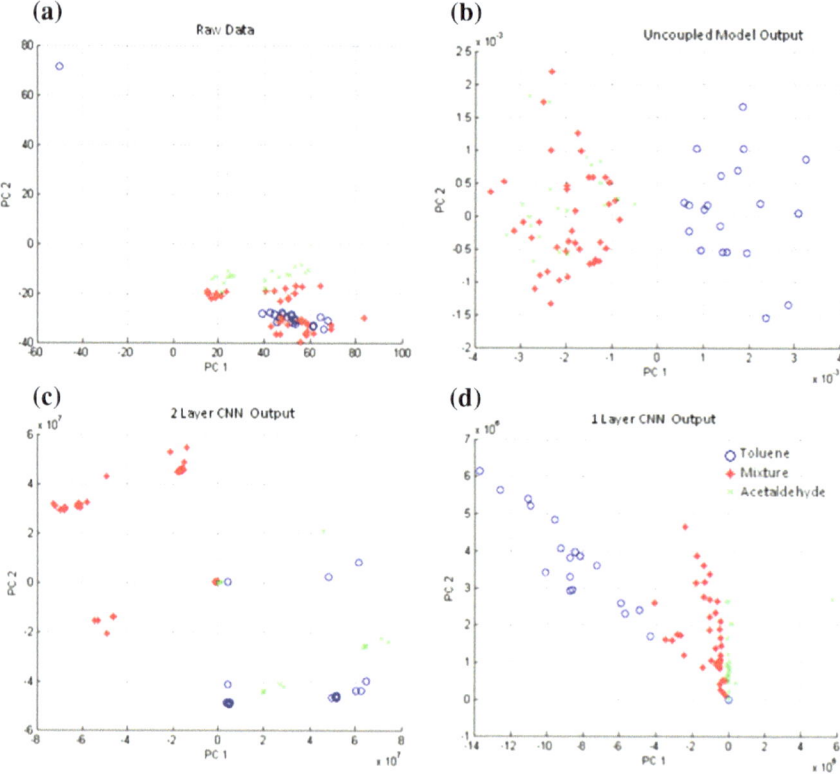

Fig. 3.7 Principal components at $T = 5$ s **a** for raw data, **b** for uncoupled model, **c** for two-layer CNN-based artificial antennal lobe, and **d** for one-layer CNN-based artificial antennal lobe

representation of mapping so that data points from different classes can be divided by a clear wide gap.

The main idea of a support vector machine is to construct a hyperplane as the decision surface in such a way that the margin of separation between positive and negative examples is maximized [18]. This aim is directed by the method of structural risk minimization of statistical learning theory. This induction principle is based on the fact that the error rate of a learning machine on test data is bounded by the sum of the training error rate and a term that depends on the *Vapnik–Chervonenkis dimension*; in the case of separable patterns, a support vector machine produces a value of zero for the first term and minimizes the second term [18].

In these experiments, a public available SVM tool, LibSVM [19] is used as classifier. As in the case for PCA, the output of the excitatory processors is collected to be classified. For each time T, projection neuron (excitatory processor) outputs are put together in a vector of length N_E and labeled with its class. For 80 records, 80 patterns that labeled with one of the three classes are obtained at every time snap. These labeled vectors are used as SVM inputs. SVM is trained with a linear kernel and leave one out technique is used for validation.

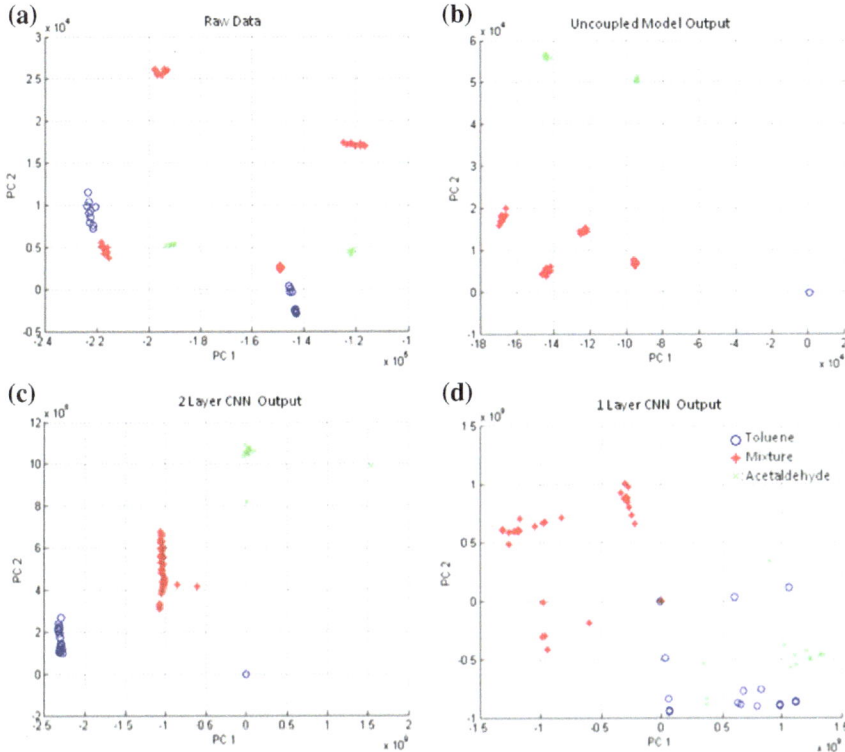

Fig. 3.8 Principal components at $T = 100$ s **a** for raw data, **b** for uncoupled model, **c** for two-layer CNN-based artificial antennal lobe, and **d** for one-layer CNN-based artificial antennal lobe

Two types of performance evaluations are done. One assuming system has memory and one considers every snapshot as a different classification problem.

3.7.3 System Without Memory

Projection neurons in the Antennal Lobe (AL) give nearly constant output to the Mushroom Body (MB) [20]. Movement of information from AL to classification layer is repeated in single time samples. The instant activity of projection neurons can be considered as the input of classification layer [20]. SVM module projects snapshots on excitatory neuron outputs. One pattern is randomly left out for test and SVM is trained with other 79 labeled patterns.

This procedure given in Fig. 3.10 is repeated for 20 times with different test patterns. If the decision of SVM is the same as the label of the test pattern, that trial is counted as a success. Overall performance at that time snapshot is [21]

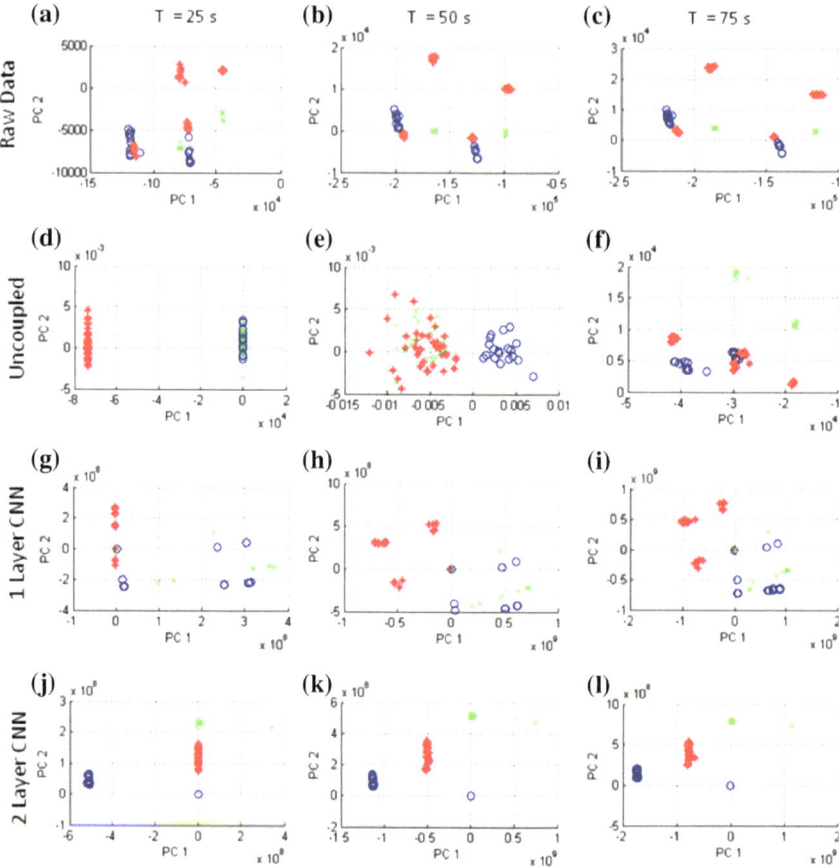

Fig. 3.9 Principal components for raw data, uncoupled model, two-layer CNN, and one-layer CNN in the rows and at $T = 25$ s, $T = 50$ s, and $T = 75$ s, in the columns

$$\text{Performance} = 100\frac{\#success}{20}\%.$$

The performance of CNN-based models and uncoupled models are given in Fig. 3.11. Estimated curves from test results when the input is started to be read at $T = 0$ till $T = 100$. Curves are fitted with polynomials where the Root Mean Square Error (RMSE) of one-layer CNN, two-layer CNN, and uncoupled curves is 10.5481, 11.5857, and 10.4164, respectively.

It can be seen that the performance of classification increases in time but the PCA results were providing better performance. Therefore, it is obvious that the classification tool used is highly important in showing the real performance of the antennal lobe model. As a comparison, the performance of two-layer CNN increases faster than one-layer CNN in small amounts up to a point. The number of connections can cause this. However, for these models, parameter search and optimization is not

Fig. 3.10 Process for testing the system without memory

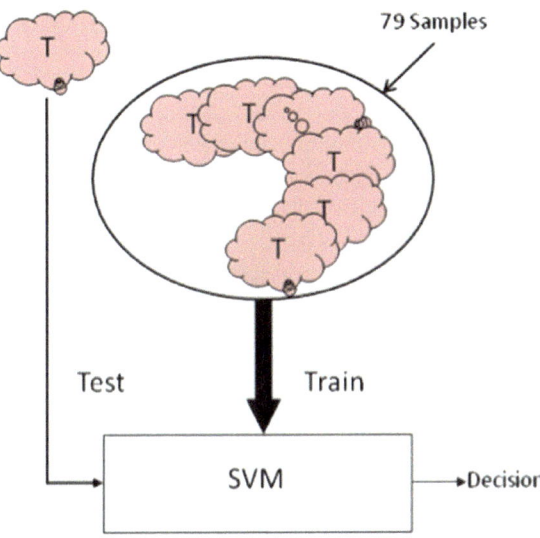

Fig. 3.11 Results for memoryless system (© 2019 IEEE. Reprinted, with permission, from Ayhan et al. [21])

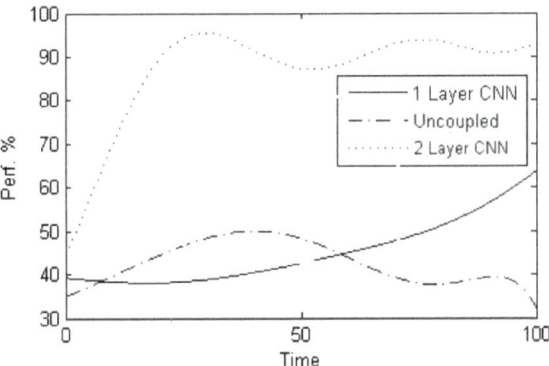

done for this data set. Therefore, a classification tool test like this SVM test does not show the true performance of the models but may give an insight about the effect of dynamics since the performance of the uncoupled model is satisfactory enough for some applications, even for this difficult classification problem. Change in parameters like gain, time constant, connection weights, and connection matrix will definitely affect the performance.

3.7.4 System with Memory

As indicated above, the classification tool or training method can affect the performance so that they may be altered in order to find the best tool and method for this problem. Support vector machine is the second tool examined with this data set, so in this subsection, the training phase of SVM is changed. In the memoryless case, all the time snapshots were considered as individual problems so distinct training sets and SVM models were built for each. However, in a real-time application, no one can be sure about a pattern belongs to which snapshot exactly. Moreover, if the system is already on up to time T, then there is no reason to dismiss the previous snapshots in decision. The system with memory is initiated with these motivations.

The introduced olfactory system with CNN-based antennal lobe is depicted in Fig. 3.12 for this case. As shown in Fig. 3.12, from time 0 up to time $T - 1$, all the patterns are saved as a training set of SVM. Then, at time T, SVM is tested by patterns of snapshot T. The performance of the model is equal to the performance of SVM at time T.

Estimated curves for test results are shown in Fig. 3.13, curves are fitted with polynomials where RMSE of one-layer CNN, two-layer CNN, and uncoupled curves are 10.8041, 11.2813, and 15.1173, respectively. CNN topologies give higher scores than uncoupled topology after the input is applied. Another point is that with this training method, a performance of nearly 80% is achieved within the first 20 s after the input is applied which is faster than the previous training case. As a result, better performance can be achieved earlier by reducing the number of connections between neurons and using a regular structure instead of a random architecture if classification method is changed as indicated in this section [21].

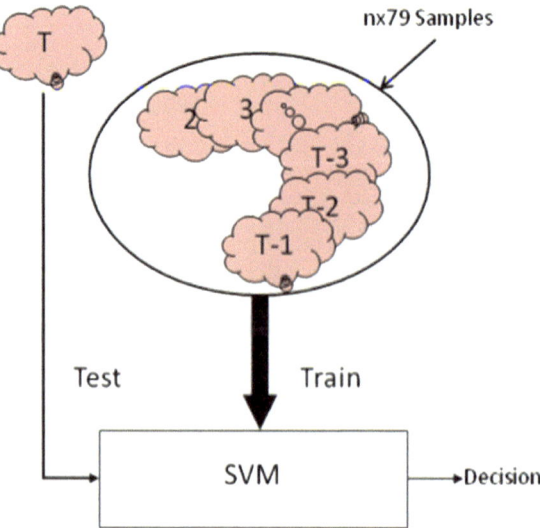

Fig. 3.12 Structure of the system with memory

Fig. 3.13 Results for system with memory (© 2019 IEEE. Reprinted, with permission, from Ayhan et al. [21])

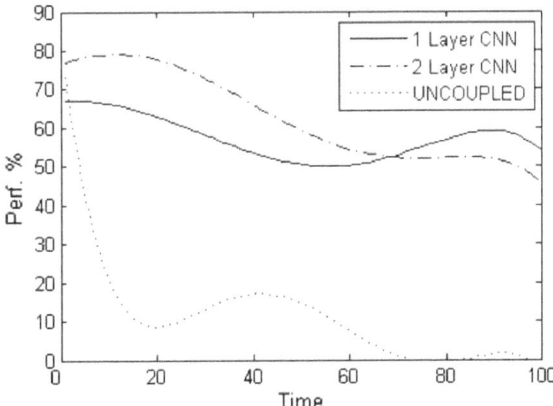

3.8 Tests for Small-World Cellular Neural Networks (SWCNNs) Based Model

Small-world CNN-based artificial antennal lobe model is tested with the data **Set B** which contains five types of odors. Small-World Cellular Neural Networks (SWC-NNs) model is introduced in Sect. 2.5. Aim for the tests is to find out the effect of small-world phenomenon on odor classification problem rather than showing the performance of another CNN-based artificial antennal lobe model with a different problem. The healing of using spatiotemporal patterns instead of spatial sensor responses is already shown with the previous tests.

In these tests, the gain in performance will be observed by increasing the number of neurons which make random connections. The classification tool is selected as SVM, and it is trained as a memoryless system, so every single snapshot is a different problem. However, leave one out technique is not used for validation. The SVM is trained with 60 of the patterns generated at time T and 15 patterns are used for the test. Then the pattern set is shuffled and this procedure is repeated ten times. The performance in the test set for each trial is recorded, and the average performance of ten trials is accepted as the performance of the SWCNN at that time T.

A two-layer SWCNN of size 4×4 is build with $S_{i,j}(1)$ and the templates

$$\hat{A}^I = \hat{A}^E = \begin{bmatrix} 1 & 1 & 1 \\ 1 & 0 & 1 \\ 1 & 1 & 1 \end{bmatrix} \tag{3.3}$$

and

$$A^I = A^E = 0, \tag{3.4}$$

which are given in Eqs. (2.29) and (2.30).

Fig. 3.14 Classification performance of SWCNN for the templates given with Eqs. (3.3) and (3.4) and $p = 0$

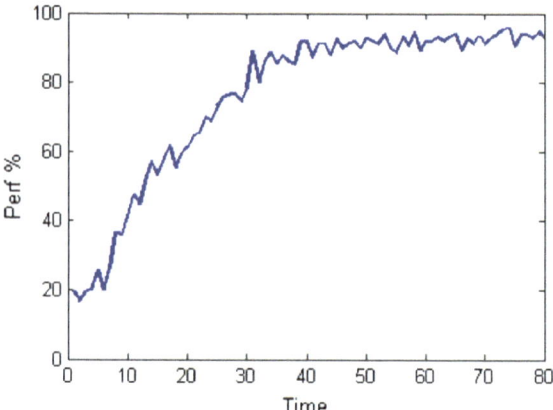

Except for the neurons at the edges, neurons have eight local couplings, and the total number of neurons in two layers is 32. The ratio of neurons that build a random connection to all neurons is given with p. All the neurons are allowed to build at most one random connection with the small-world map $M(i, j; m, n)$ which is defined in Sect. 2.5. Here N_c is defined as the number of excitatory neurons, which builds a random connection, and p is calculated as

$$p = \frac{N_c}{n \times m}.$$

One should note that connections between the same type of neurons are forbidden. The value of p changes here between 0 and 1.

In Fig. 3.14, increasing performance of the network designed as a basis for SWCNN can be monitored. The question is whether the number of random connections can increase the performance or not. Each of the cells randomly chosen from the excitatory layer made a short cut to an inhibitory neuron which it is not locally coupled. If all the neurons in the excitatory layer have one random connection, then N_c is equal to 16 hence $p = 1$.

The performance test is applied to those 16 cases for five times: the edges of random connections are randomly changed, and then the average performance is recorded. All those 16 cases are given in Fig. 3.15. The earliest time that the system achieves its maximum performance is not changed much.

The network is driven by 16 sensors in the cases above but can the increasing number of random connections compensate loss of sensor data in the sensor array? To answer this question, the cells in the network are driven by the same sensor. First, the performance of the network with only local couplings is given in Fig. 3.16. When Figs. 3.16 and 3.14 are compared, the maximum classification performance is decreased when the sensor array data is lost.

The number of random connections is limited to 16 in the previous scenarios, a more clear observation is tried to be done by enlarging the network layers to a size of

Fig. 3.15 Classification
performance of SWCNN for
the templates (3.3) and (3.4)
and $1 \leq p \leq 16$

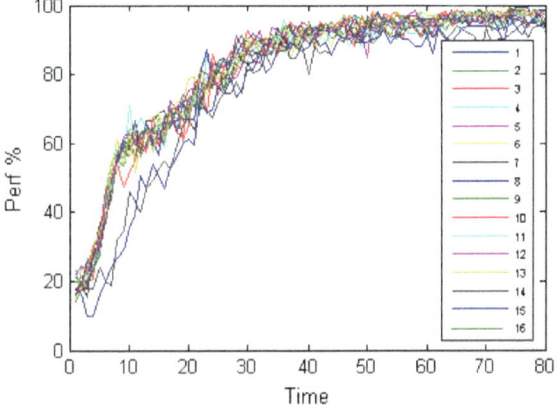

Fig. 3.16 Classification
performance of SWCNN for
the templates (3.3) and (3.4)
and $p = 0$ driven by one
sensor

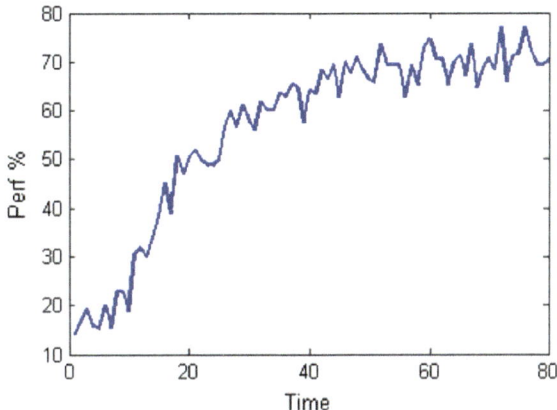

Fig. 3.17 Classification
performance of 128 cell
SWCNN for the templates
(3.3) and (3.4) and
$1 \leq p \leq 16$

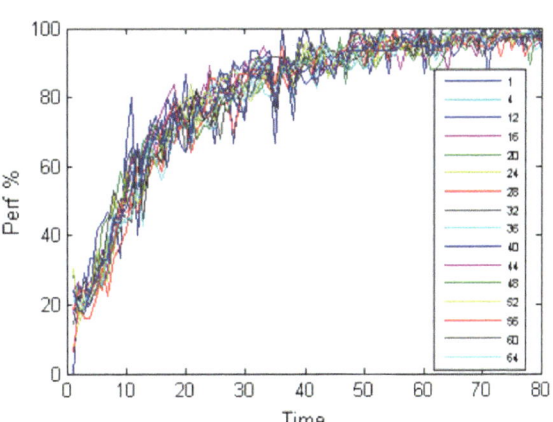

8×8 and increasing the number of sensors to 16 again. The performance change is given in Fig. 3.17 changing p from 1 to 64 in steps of four for clarity. In conclusion, SWCNN does not improve classification performance and times when it is compared with the other CNN-based model. However, it is reported in [22] that SWCNN is more fault tolerant than the other CNN-based models.

3.9   Tests for Reconfigurable Cellular Neural Network (RCNN) Based Model

In a metal oxide gas sensor, the odor in the environment changes the connectivity of the sensor surface. This change can be measured and collected with a simple electrical circuit. The time required for a sensor to settle on a stable value is nearly 80 s. These odors can be classified with the stable values of the sensors but it may be too long for a mobile application such as robot navigation. Therefore, not only to classify the odors precisely but also classify them by transient response of the sensors is important. The aim here is to accelerate the odor classification by using a CNN-based artificial antennal lobe for feature extraction. The procedure is summarized in Fig. 3.18.

A 2-D CNN layer of 5×5 RCNN which was introduced in Sect. 2.6.1 is created for this problem set. The H matrix is randomly formed by MATLAB. Another randomly formed matrix B indicates the match between sensors and the cells. All the cells receive one and only one sensor output such that all the sensors are connected at least one cell. The network is run and the output of the cells which are labeled as excitatory by the H matrix are collected. Each time snapshot is considered as a different classification problem. Therefore, there are 80 different problems regarding 80 s of the records. For every time snapshot T, the features of size $n_{ex}1$ are formed with the output of the excitatory neurons at that time. Principal component analysis is done on the features for classification. The aim is to achieve a distinguishable feature set as quickly as possible. However, the characteristics of the features change due to the topology used in the antennal lobe model. There is no key evidence for the best topology for a given problem set, so the optimum topology should be found through trials. The H matrix is changed randomly and the procedure is repeated. Results are given in appendix for three different H matrices to show the effect of topology on discrimination.

In Figs. 3.19, 3.20 and 3.21, only first 10 s of the excitatory neuron outputs are considered. In the subfigures, the topology created with the random H matrix and the first two principal components of the features is given. As it is clearly seen from the figures, the use of antennal lobe may accelerate the classification but H matrix, given in the figures, changes the property of neuron output. A parameter search on connection weights, time constant, and input gain is required for the best classification performance.

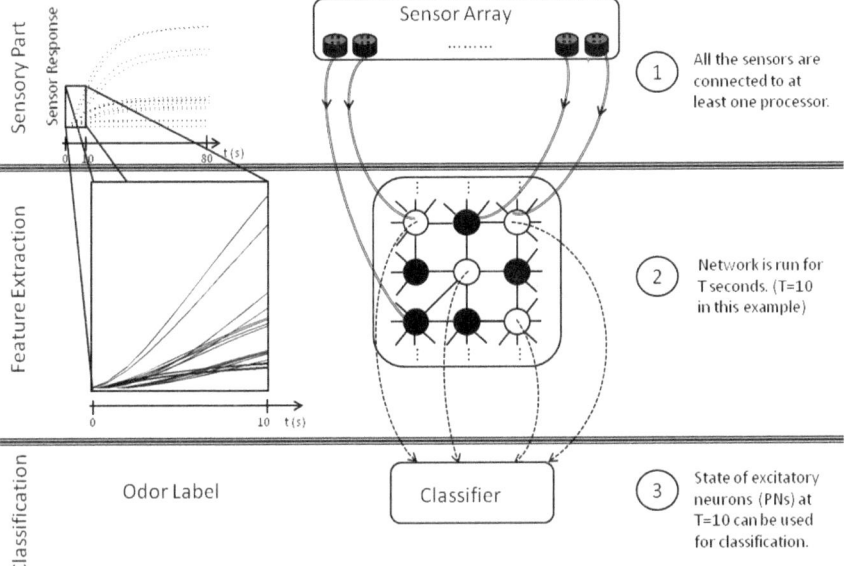

Fig. 3.18 Summary of evaluation setup. Odor data is collected by the sensor array which contains 16 Figaro Gas Sensors. All the sensors are connected to at least one processor in the antennal lobe model which is built according to the proposed CNN structure. The response of the sensor array to acetone (200 ppm) is also plotted. As seen from the sensor response, one of the challenges in odor classification problem is to extract information from the early stages in the sensor transient. In the figure, the network is fed only with the first 10 s sensory data. Then the states of excitatory neurons which stand for the projection neurons in the biological model can be used for classification

In the network, connections are built in sphere of influence $S_{i,j}(r = 1)$ and the connection template is given by

$$
A = \begin{bmatrix} 1 & \frac{1}{2} & 1 \\ \frac{1}{8} & 0 & \frac{1}{8} \\ 1 & \frac{1}{4} & 1 \end{bmatrix}.
\tag{3.5}
$$

Other constants such as β, K and g_{inp} in Eqs. (2.36) and (2.35) are chosen among some trials as 0.0156, 1, and 0.0001, respectively.

Differential equations are solved with MATLAB by Runge–Kutta method. As the sensors become stable, the separability of the classes is expected to be increasing. However, without any spatiotemporal coding, achieving an adequate separability in 10 s is not likely. Using the proposed model with the motivation of accelerating the separability, a time interval of 10 s is evaluated. When the Figs. 3.19, 3.20 and 3.21 are examined, we can conclude that different topologies move the principal components of classes in different directions. For example, in Fig. 3.21, all classes are linearly separable with two principal components at $T = 10$ where in Figs. 3.19 and 3.20, butane and ethylene are convoluted.

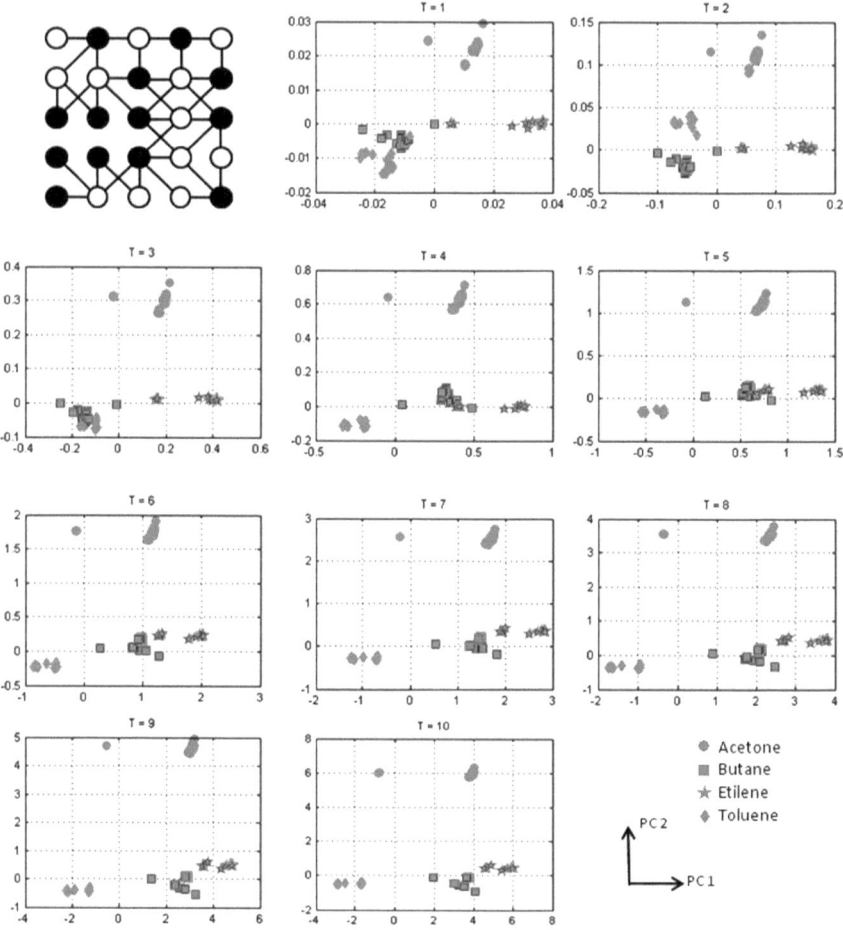

Fig. 3.19 Principal components of feature extractor output at time = {1, 2, ..., 10}—topology H1

3.9.1 EEG Signal Classification

Similar feature extraction can be done for EEG signal classification problem. For classification of EEG signals, time domain, frequency domain, or mixed features are used. CNN is recently used in EEG signal processing for signal identification and epileptic seizure prediction [23]. There are various ways of feature extraction for EEG signal processing [24]. These methods include transforms such as wavelet, discrete Fourier, discrete cosine; statistical methods like correlation between electrodes, mean, variance, skewness, and kurtosis inside a time segment; other time-domain properties of the signals like maximum and minimum value in the time segment. However, a feature that fits all problems is not valid. Every problem using EEG signal as system input from the classification of motor tasks to epilepsy seizure pre-

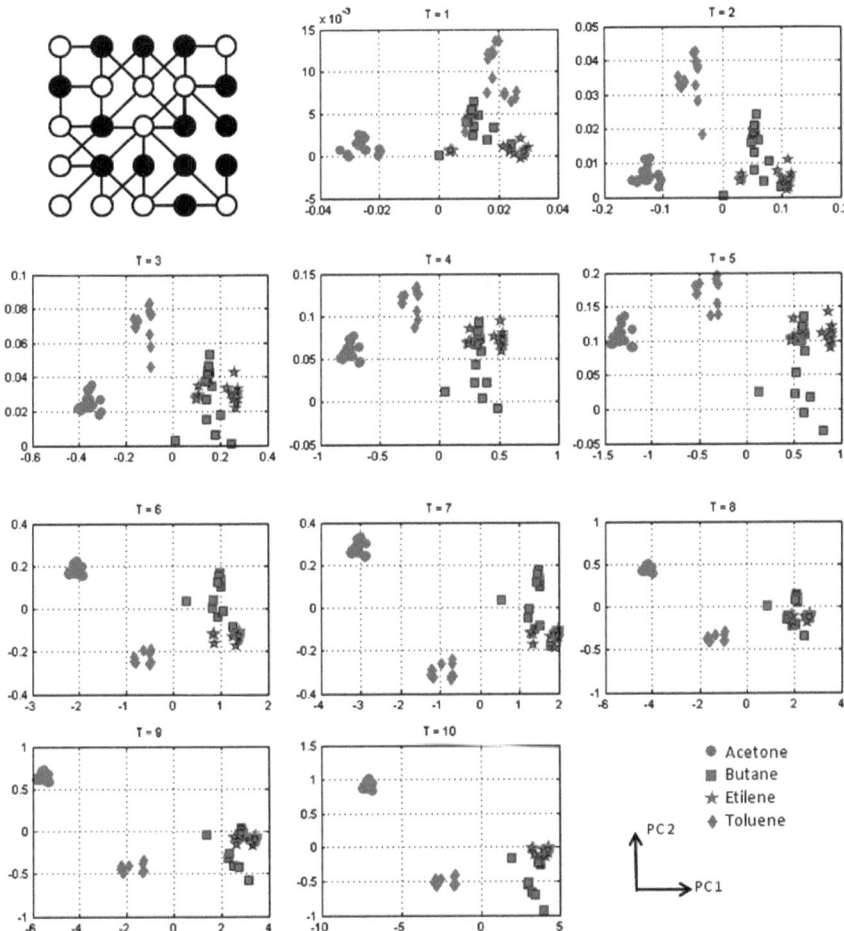

Fig. 3.20 Principal components of feature extractor output at time = {1, 2, ..., 10}—topology H2

diction requires different features for the best performance. Moreover, for the same problem, the EEG signals recorded from different people or from the same person at different times do not show the same characteristics. Therefore, a generic feature extractor is necessary. A reconfigurable cellular neural network is suitable for finding the best features of EEG data for a specific problem.

Here, EEG signals are classified in order to navigate a mobile robot. Two volunteers are asked to think on moving the robot to right, left, and straight. EEG signals are collected with a headset of 13 electrodes, namely, EMOTIV® [25], with the volunteers written permission. The data is considered as segments of 1 s recording which means 128 samples. The estimated bias value of each sensor is subtracted and the data is scaled.

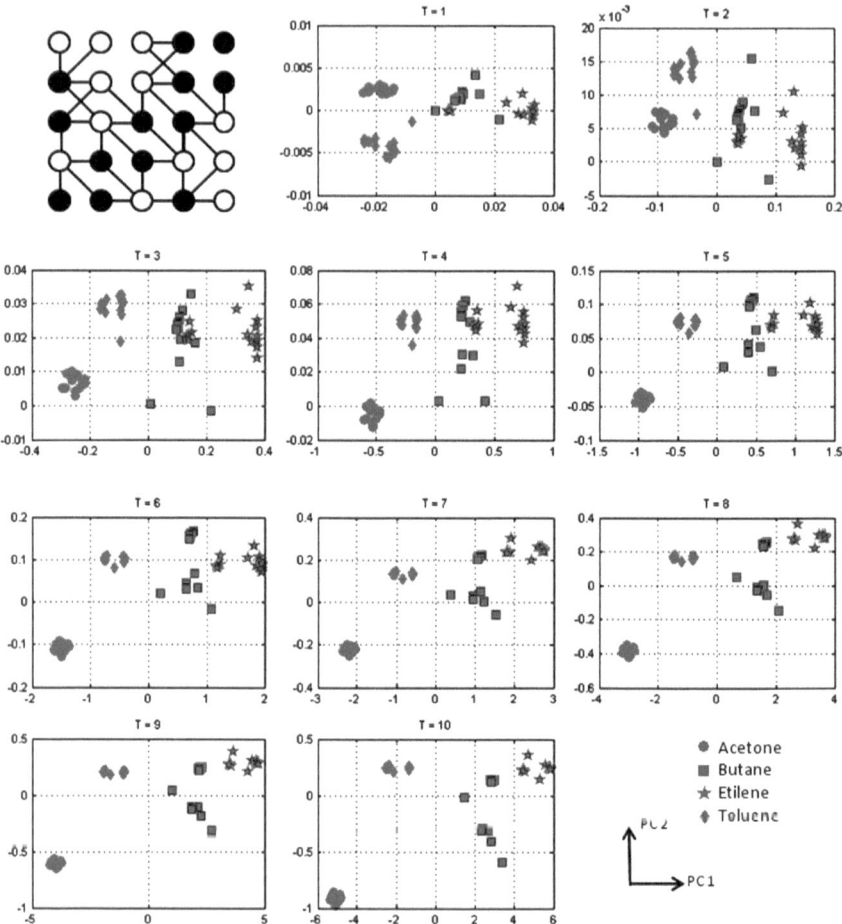

Fig. 3.21 Principal components of feature extractor output at time = {1, 2, ..., 10}—topology H3

As in the odor classification case, the H matrix is formed randomly for a 5 × 5 2-D RCNN network. The connections are determined such that the network curves on itself. In other words, there is no boundary condition for the network because the cells at the edges are connected to the cells on the other edge according to the template. The template and other constants are the same as in the odor classification problem. The procedure is also same except, the network is fed by electrode responses gathered from EMOTIV headset for 1 s duration with a sampling frequency of 128 samples per second. The CNN is run with the EEG signals of 1 s duration. The feature generated for the corresponding segment is the last value of the excitatory neurons. The features collected from the excitatory neurons are applied to principal component analysis as in the odor classification case. The first and the second principal components of the features are plotted in Fig. 3.22. As it is clear, different topologies generate

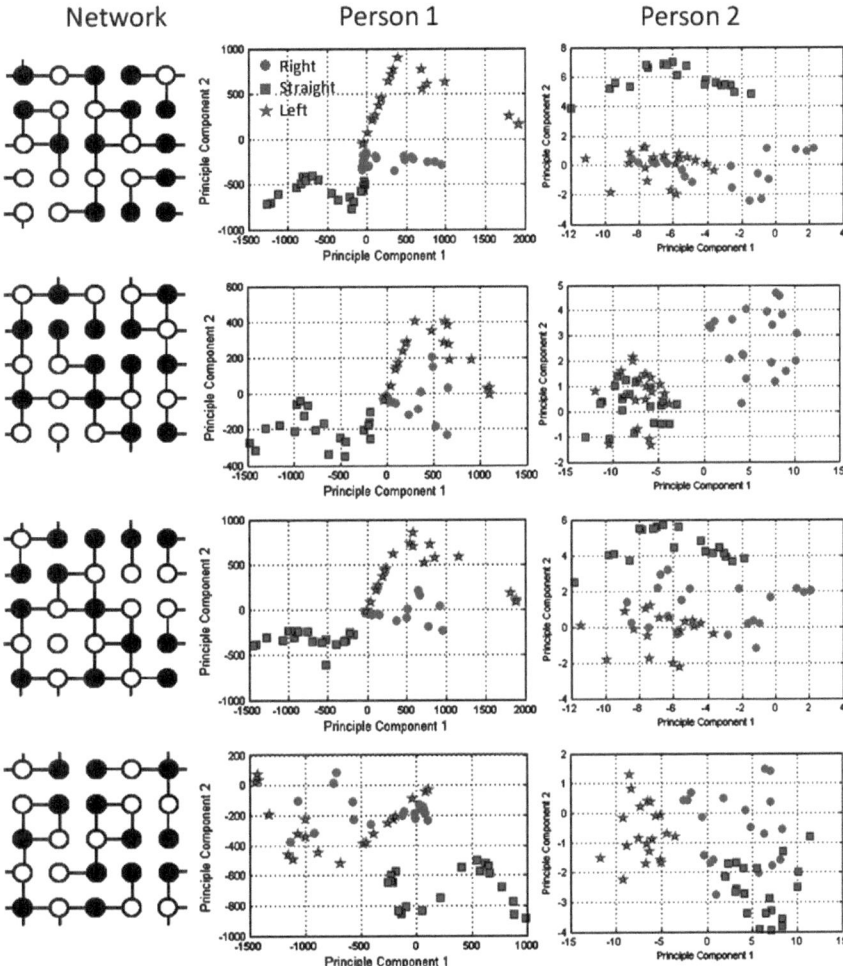

Fig. 3.22 EEG signal classification with RCNN. The first column which is named network shows the randomly located exhibitory (white) and inhibitory (black) neurons in 5 × 5 network. Second and third columns are the first and the second principal components of the features of two volunteers' EEG signals for the corresponding network topology given in the first column. This test indicates that topological diversity of the network assists to obtain separable features for a difficult classification problem

features of different separability for two volunteers because thinking methods for those volunteers are not identical. For example, it can be seen that the network on the top generates features that are linearly separable for the data collected from the first and second volunteers; however, when the features are obtained with the network on the bottom, the two principal components of the features are not linearly separable anymore. This fact emphasizes the need for reconfigurable devices for an EEG signal-driven system.

3.10   The Role of Network Topology

The reconfigurable cellular neural network is regular due to its connection template and contains randomness due to the location of subpopulations. Randomly generated ID matrix that defines which subpopulation is located in the spatial space. In fact, ID matrix describes network topology. For instance, different network topologies have been used in the previous section, and network topologies are shown in the first column of Fig. 3.22.

Here, the olfactory sensor processing application is reconsidered in order to understand the performance of randomly built topologies. For an $M \times N$ grid, which includes N_{ex} excitatory and N_{inh} processors, the number of possible different topologies can be calculated with

$$\frac{(N_{ex} + N_{inh})!}{N_{ex}! \times N_{inh}!}. \tag{3.6}$$

An 6×6 regular CNN grid is used for feature extraction, where randomly determined H matrix, which is composed of ones and zeros, locates the excitatory and inhibitory neurons (ID matrix H is defined in Sect. 2.6.1). All the neurons receive one and only one external input from a sensor provided that all the sensors are connected at least one neuron. The weight of the sensor input to the neuron (g_{inp}) is 1 for all the neurons. The feature extractor is run for the first 15 s of the record with the following constants: $\beta = \frac{1}{64}$, $K = 1$, $g_{inp} = 0.0001$ and

$$A = \begin{bmatrix} 0 & \frac{1}{2} & 0 \\ \frac{1}{8} & 0 & \frac{1}{4} \\ 0 & \frac{1}{8} & 0 \end{bmatrix}, \tag{3.7}$$

which are defined in Eqs. (2.35) and (2.36).

The last states of the excitatory neurons form the feature for the related record. Therefore, features of size 18 are created for each of the records belonging to each class. Then k-means clustering is applied to the excitatory neuron outputs. Then the total number of misclassified points is recorded for each data set. This procedure is repeated for 300 topologies shown by T_i, which are generated by randomly chosen H matrices. The best topology is tried to be found from 300 trials.

The effect of diversity of network topology is tested on odor classification problem and the measurement which is given in Sect. 3.6 is also used here. For the first set (**Set A**), classification performance among 300 topologies changes between 49 and 87%. We can obtain the highest result (87%) when we use one of these fifteen topologies:

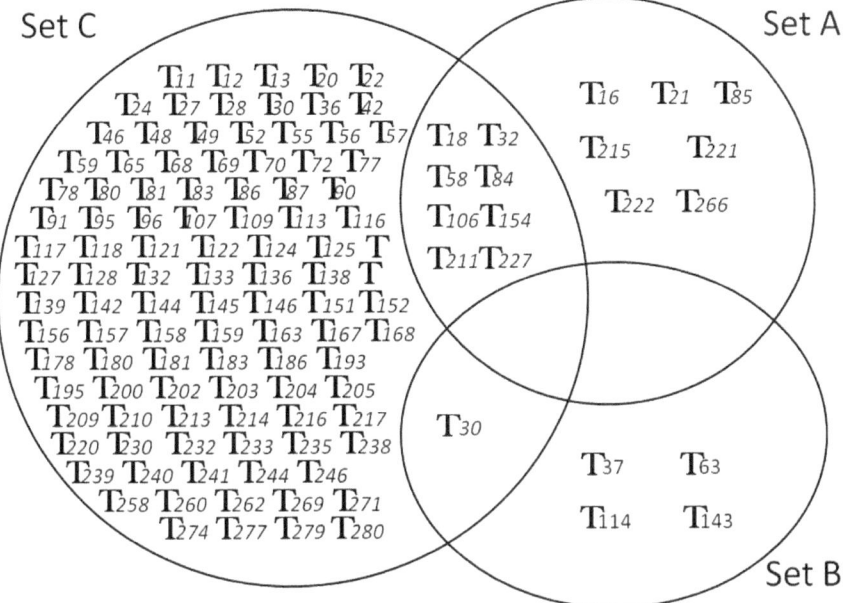

Fig. 3.23 300 different topologies are built randomly by changing ID matrix H and named T_i. They are used as feature extractors in three different data sets: **Set A**, **Set B**, and **Set C** (Given in Sect. 3.6). The classification performance for every trial is recorded. The T matrices that give the highest classification performance are pinned inside the Venn diagram of the corresponding set or inside the intersection. None of those 300 topologies give the best performance for both of the three sets (© 2019 IEEE. Reprinted, with permission, from Ayhan, T. and Yalcin, M.E.: Randomly Reconfigurable Cellular Neural Network. in Proceedings of the 20th European Conference on Circuit Theory and Design (ECCTD11), 625–628 (2011))

T_{16}, T_{18},T_{21}, T_{58}, T_{84}, T_{85},T_{106}, T_{154}, T_{211}, T_{215},T_{221}, T_{222}, T_{227} and T_{266}. These topologies are written inside the circle **Set A** in Fig. 3.23. Similarly, the classification performance on **Set B** varies from 60 to 98%. Topologies resulting a classification performance of 98% are given in Fig. 3.23. The classification performance on **Set C** changes between 67 and 98%. Best topologies are given in Fig. 3.23.

Briefly, if a successful feature extractor topology is found for a problem set containing three classes, there is no strong evidence that it will result in a similar performance even one of the odors in the set is common in the two sets. The classification performances highly depend on the feature extractor. By trying different topologies randomly, we can find a topology that results in high performance. The advantage is that we can search though topologies by changing only the ID matrix not the entire cells in the network. The conclusion of these tests indicates that a single network topology does not exist to perform well for different problems.

References

1. T.C. Pearce, S.S. Schiffman, *Handbook of Machine Olfaction: Electronic Nose Technology* (Wiley-VCH, 2003)
2. P. Nef, How we smell: the molecular and cellular bases of olfaction. News Physiol. Sci. **13**, 1–5 (1998)
3. M. Zarzo, The sense of smell: molecular basis of odorant recognition. Biol. Rev. **82**(3), 455–479 (2007)
4. L. Buck, R. Axel, A novel multigene family may encode odorant receptors—a molecular-basis for odor recognition. Cell **65**(1), 175–187 (1991)
5. Q. Gao, B. Yuan, A. Chess, Convergent projections of Drosophila olfactory neurons to specific glomeruli in the antennal lobe. Nat. Neurosci. **3**(8), 780–785 (2000)
6. A.F. Silbering, C.G. Galizia, Processing of odor mixtures in the Drosophila antennal lobe reveals both global inhibition and glomerulus-specific interactions. J. Neurosci. **27**(44), 11966–11977 (2007)
7. B. Hansson, Olfaction in lepidoptera. Experientia **51**(11), 1003–1027 (1995)
8. N. Strausfeld, L. Hansen, Y. Li, R. Gomez, K. Ito, Evolution, discovery, and interpretations of arthropod mushroom bodies. Learn. Mem. **5**(1–2), 11–37 (1998)
9. W. Zhou, D. Chen, Binaral rivalry between the nostrils and in the cortex. Curr. Biol. **19**(18), 1561–1565 (2009)
10. E. Morrison, R. Costanzo, Morphology of olfactory epithelium in humans and other vertebrates. Microsc. Res. Tech. **23**(1), 49–61 (1992)
11. H. Hotelling, Analysis of a complex of statistical variables into principal components. J. Educ. Psychol. **24**, 417–441 (1933)
12. C.E. Shannon, A mathematical theory of communication. Bell Syst. Tech. J. **27**, 379–423 (1948)
13. A. Vergara, S. Vembu, T. Ayhan, M.A. Ryan, M.L. Homer, R. Huerta, Chemical gas sensor drift compensation using classifier ensembles. Sens. Actuators B Chem. **166167**, 320–329 (2012)
14. Figaro USA, Inc., http://www.figaro.co.jp/en/. Accesssed 15 Feb 2019
15. A. Setkus, A. Olekas, D. Senuliene, M. Falasconi, M. Pardo, G. Sberveglieri, Featuring of odor by metal oxide sensor response to varying gas mixture, in *Olfaction and Electronic Nose, Proceedings, of AIP Conference Proceedings*, ed. by M. Pardo, G. Sberveglieri, vol. 1137 (2009), pp. 202–205
16. K. Pearson, On lines and planes of closest to systems of points in space. Philos. Mag. **2**(6), 559–572 (1901)
17. V. Vapnik, *The Nature of Statistical Learning Theory* (Springer, N.Y., 1995)
18. S. Haykin, *Neural Networks: A Comprehensive Foundation* (Prentice Hall, 1998)
19. C.C. Chang, C.J. Lin, LIBSVM: a library for support vector machines, v2.85, software available at https://www.csie.ntu.edu.tw/~cjlin/libsvm/. Accessed 15 Feb 2019
20. M.K. Muezzinoglu, A. Vergara, R. Huerta, T. Nowotny, N. Rulkov, H.D.I. Abarbanel, A.I. Selverston, M.I. Rabinovich, Artificial olfactory brain for mixture identification, in *NIPS* ed by D. Koller, D. Schuurmans, Y. Bengio, L. Bottou, (MIT Press, 2009), pp. 1121–1128
21. T. Ayhan, K. Muezzinoglu, M.E. Yalcin, Cellular neural network based artificial antennal lobe, in *Proceedings of the 12th IEEE International Workshop on Cellular Neural Networks and their Applications (CNNA 2010)* (2010), pp. 1–6
22. T. Ayhan, M.E. Yalcin, An application of small-world cellular neural networks on odor classification. Int. J. Bifurc. Chaos **22**(1), 1–12 (2012)
23. F. Gollas, C. Niederhoefer, R. Tetzlaff, Toward an autonomous platform for spatio-temporal EEG signal analysis based on cellular nonlinear networks. Int. J. Circuit Theory Appl. **36**(10), 623–639 (2008)
24. S. Sanei, J. Chambers, *EEG Signal Processing* (Wiley-Interscience, 2007)
25. EMOTIV, http://www.emotiv.com/. Accessed 15 Feb 2019

Chapter 4
Implementations of CNNs

4.1 Introduction

Today, many applications demand high-speed performance to achieve real-time operation. Field Programmable Gate Array (FPGA) devices which are an implementation alternative to have high-speed performance on computationally intensive tasks exploit hardware parallelism. FPGAs allow circuit designers to produce application-specific chips bypassing the time-consuming and expensive fabrication process. They are composed of three fundamental components: logic cells, I/O blocks, and programmable routing. The logic cell structure typically consists of lookup tables (LUTs), carry logic, flip-flops, and programmable multiplexers. The I/O blocks provide FPGAs to interact with the external world. The programmable routing is configured to make all the required connections between logic blocks. An outstanding growth of resources in FPGA devices utilizes the implementation of a full application in a single chip. Furthermore, merging programmable logic and a microcontroller unit in the same chip converts the FPGA into a System-on-Chip (SoC) device. An FPGA device could bring an increase in performance compared to a sequential implementation of processing on the general-purpose processor, if architecture for the processing is designed with the right technique.

Processing of two-dimensional data such as image processing and spatiotemporal active wave generation is well suited for exploiting parallelism; therefore, the use of an FPGA device can bring an increase in performance compared to general-purpose processor. Furthermore, converting FPGA chip into a System-on-Chip (SoC) device allows to allocate the computationally intensive tasks of any target application to digital hardware, and the remaining processing tasks exhibit less parallelism to embedded software which is executed on the microcontroller.

Cellular nonlinear network has been introduced as a special high-speed parallel neural structure for array processing application such as image processing and recognition. Emulation of CNN's behavior on the general-purpose processor is in fact limited computation power of its array processing abilities. In this chapter, hardware accelerators for implementations of CNN will be presented. The architectural design of accelerators on FPGAs aims to combine the full advantage of massive parallel computing power of FPGA and parallel structure of CNN.

M. E. Yalçın et al., *Reconfigurable Cellular Neural Networks and Their Applications*, SpringerBriefs in Nonlinear Circuits, https://doi.org/10.1007/978-3-030-17840-6_4

We first present a hardware accelerator implemented in FPGA which performs solving nonlinear differential equations of the CNN given in Sect. 2.3.2 to obtain a variety of spatial–temporal dynamical behavior. Handling the computational complexity of differential equations on the accelerator inspires to exploit the dynamic behavior which is obtained from these equations in applications. In this chapter, an algorithm to solve the robot path finding problem by observing the motion of the wave front of the active waves computed on the accelerator is introduced. Second hardware accelerator implemented in FPGA aims to solve the speed problem in artificial olfaction systems. The accelerator is functioning like the reconfigurable cellular neural network which was introduced in Sect. 3.9 together with its configurability. Our solution provides reconfiguration of a large network with different types of processors so that the network can be adapted to many different problems or can be used to mimic different metabolic functions. Beside the accelerator, an SoC design is introduced in this chapter to achieve complete artificial olfaction system on an FPGA.

4.2 Digital Implementation and Application of Cellular Nonlinear Network

4.2.1 Digital Implementation of Locally Coupled Oscillatory Network

Locally Coupled Oscillatory Network (LCON) was already introduced in Sect. 2.3.2. In this section, LCON is used for generating spatiotemporal pattern emulating the network on an FPGA. The wave computing system which includes host computer, the FPGA development board, and monitor is illustrated in Fig. 4.1. The forward Euler integration of the LCON which was given Eq. (2.17) is

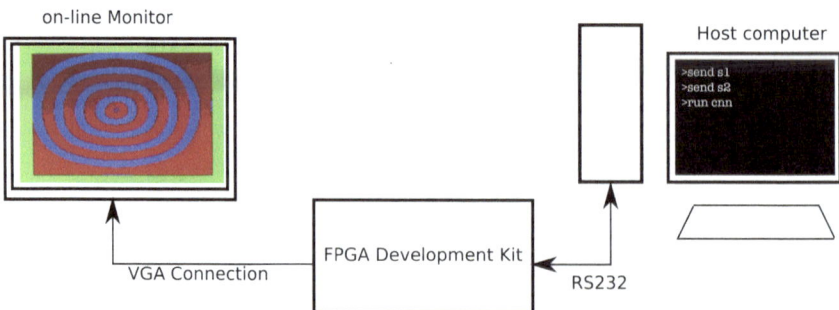

Fig. 4.1 The scheme of the wave computing system. Initial values of cells, control signals, and parameters which are given in Table 4.1 from the host computer are transferred to the FPGA development board. The waves which are generated on the FPGA development board are simultaneously monitored on a VGA screen

$$x_{i,j}(k+1) = x_{i,j}(k) + \tau[\alpha x_{i,j}(k) + \beta y_{i,j}(k) + g(x_{i,j}(k)) + w\Im_{i,j}(k) + U_{i,j}],$$
$$y_{i,j}(k+1) = y_{i,j}(k) + \tau[\gamma x_{i,j}(k) + \epsilon y_{i,j}(k)], \tag{4.1}$$

with the nonlinearity,

$$g(x_{i,j}(k)) = \begin{cases} m \cdot (x_{i,j}(k) - \lambda) & \text{if } x_{i,j}(k) > \lambda; \\ 0 & \text{if } |x_{i,j}(k)| \le \lambda; \\ m \cdot (x_{i,j}(k) + \lambda) & \text{if } x_{i,j}(k) < -\lambda; \end{cases} \tag{4.2}$$

and the synaptic law,

$$\Im_{i,j}(k) = a_{i,j+1}x_{i,j+1}(k) + a_{i-1,j}x_{i-1,j}(k) + a_{i,j-1}x_{i,j-1}(k) + a_{i+1,j}x_{i+1,j}(k), \tag{4.3}$$

which defines the effect of coupled neighbors of the node.

Nodal Processing Element (NPE) is the principal component of this digital design. NPEs compute each cell states solving the mathematical Eq. (4.1). In our design, only 16 NPEs concurrently execute iterations which means that the state values of 4×4 locally coupled cells are simultaneously computed from Eq. (4.1). In this design, these 16 NPEs constitute the Cellular Nonlinear Processor Network (CNPN).

Parameters for the mathematical model (4.1) are transferred from parameter register to NPE. Each NPE has two stacks called s_1 and s_2, and five temporary registers called f_1, f_2, a_1, a_2, and t, as shown in Fig. 4.2 and Table 4.1.

In Table 4.1, α, β, ϵ, and σ are the state coefficients, τ is the integration time step, a is the coupling coefficient, m is the slope in nonlinearity, and u is the input. The discontinuity point on nonlinearity (λ) is fixed to 1 and is not loaded to the stacks.

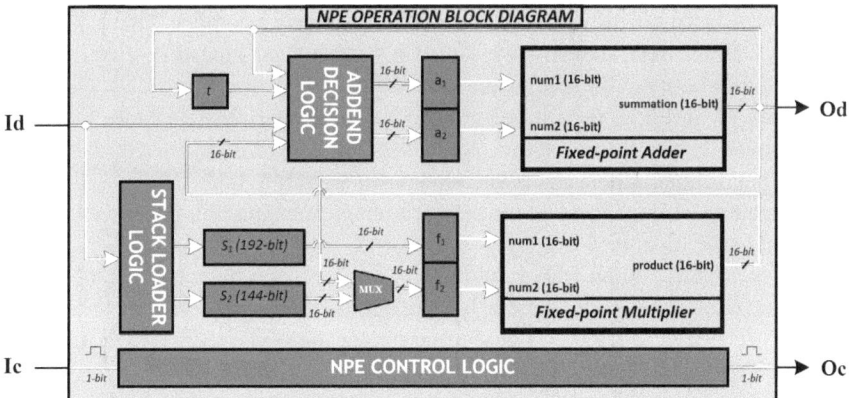

Fig. 4.2 Block diagram (© 2019 IEEE. Reprinted, with permission, from Yeniceri and Yalcin [1]) of the Nodal Processing Element (NPE), **Id** variable and parameter inputs, **Od** variable outputs, **Ic** control signal inputs, and **Oc** control signal outputs

Table 4.1 Parameters and state variables hold by s_1 and s_2 stacks in the NPE

Parameters and variables	s_1 (192-bit)	s_2 (144-bit)
Word 1	$a_{i,j+1}$	$x_{i,j+1}(k)$ or x_{fixed}
Word 2	$a_{i,j-1}$	$x_{i,j-1}(k)$ or x_{fixed}
Word 3	$a_{i-1,j}$	$x_{i-1,j}(k)$ or x_{fixed}
Word 4	$a_{i+1,j}$	$x_{i+1,j}(k)$ or x_{fixed}
Word 5	α	$x_{i,j}(k)$
Word 6	β	$y_{i,j}(k)$
Word 7	m	$u_{i,j}$
Word 8	τ	$x_{i,j}(k)$
Word 9	1	$y_{i,j}(k)$
Word 10	ϵ	
Word 11	σ	
Word 12	τ	

Wave computer core includes clock generator, control circuit, parameter register, CNPN circuit, CNPN cache, and communication interfaces as shown in Fig. 4.3. The wave computer core is designed, implemented, and programmed by Xilinx® ISE Design Suite. When the core is programmed, the initial values of nodes, control signals, and parameters are transferred to the core by using built-in communication interfaces and MATLAB® functions.

All types of data which are required by the wave computer are sent by the host computer to the control block. Control block receives the data and distributes it to the memory block and parameter register. As shown in Table 4.1, s_1 and s_2 stacks are loaded with 12 and 9 words, respectively, by stack loader logic in NPE circuit. Each word is 16-bit in width and stores the fixed-point number. These words are used by CNPN circuit sequentially. The data that belongs to network slices is carried continuously between the CNPN and the memory block, during the CNN emulation. Each one of 16 NPEs uses its parameters and state variables in order to execute iteration for its own x and y states. When the slowest NPE completes its calculation, CNPN control circuitry sends acknowledgment signals to make NPEs ready for the next iteration. Simultaneously, the network image is captured by observation block and sent to the monitor.

State variables, initial values, and parameters of the system are 16-bits in width and have $Q7.9$ fixed-point number format. As the represented numbers are signed, format is changed to $Q6.9$ as 1 bit is reserved for the sign:

$$x_{[\text{B10}]} = \frac{1}{2^n}\left[-2^{N-1}b_{\text{N}-1} + \sum_{i=0}^{N-2} 2^i b_i\right]. \tag{4.4}$$

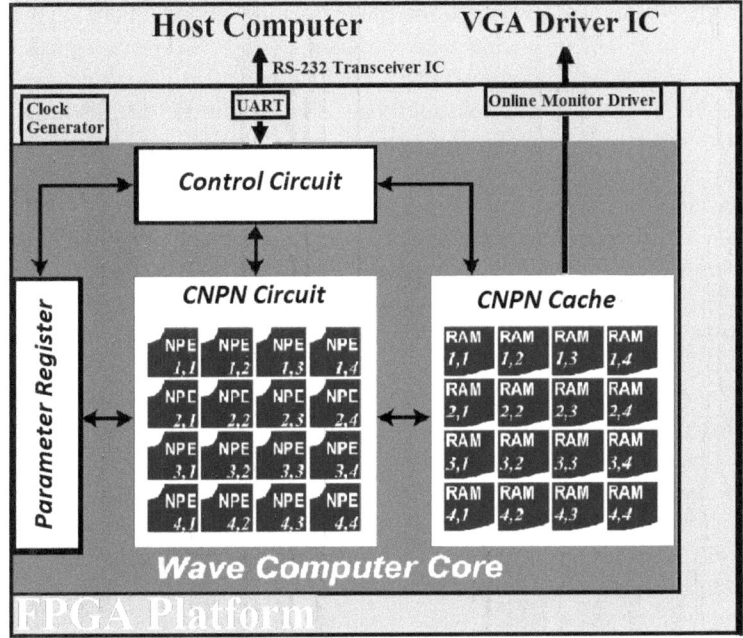

Fig. 4.3 Block diagram of the wave computer core (© 2019 IEEE. Reprinted, with permission, from Yeniceri and Yalcin [1])

In Eq. (4.4), $x_{[B9]}$ represents the fractional number in decimal, $x_{min} = -2^m$, $x_{max} = 2^m - 2^{-n}$, and $resolution = 2^{-n}$. It can be observed that $Q6.9$ signed two's complement format gives us $[-64, 63.99805]$ value range and $1.953 \cdot 10^{-3}$ resolution which are used in NPE operations during CNN emulation.

Three peripheral circuits are used in the wave computer core design as shown in Fig. 4.3. Clock generator generates the low-frequency clock signal required by the Core. In this design, the whole implementation covers the 64% of the FPGA chip better than previous work which covers 76% of the FPGA chip [1]. Routed signal paths have high delays for onboard 100 MHz clock signal; therefore, clock frequency is divided into 4 to get 25 MHz by a frequency divider circuit. According to the timing analysis, the maximum clock frequency is 46,70 MHz which has been recorded 35.85 MHz for the design in [1]. Also, in order to observe the wave evolution, simultaneously the online monitor is used [1].

CNPN Cache needs 160 kB of total memory required by emulation of 16,384 nodes. No external memory is needed as FPGA chip has 272 kB BlockRAM. 16 BlockRAM modules each having 10 kB capacity are defined for this implementation. Words at different addresses can be accessible at the same time because each BlockRAM has a dual-port interface. A-ports of BlockRAMs are used to read and write data that CNPN needs, while B-ports are used for reading the data to depict the network image onto the online monitor.

Each NPE circuitry has two subcircuits which are adder circuit and multiplier circuit. These arithmetic circuits work with `permission_input` signals coming from related NPE and operate addition and multiplication operations combinatorially. After addition or multiplication operation, arithmetic circuits send control signals in order to indicate that circuits are ready for new calculation. Adder circuit and multiplier circuit are designed specially for this fixed-point arithmetic Core implementation.

The whole design is implemented on a Xilinx XC2VP30-FF896 FPGA chip. When the whole system operation is examined, the input image is scaled to a 128×128 matrix by a MATLAB script at the beginning. Each cell that is equal to one pixel on the network image functions as a relaxation oscillator using the mathematical model in Wave Computer Core design. FPGA is programmed by using Xilinx IMPACT interface, and initial values of the network image, parameters, and control signals are transferred from MATLAB to FPGA chip through RS-232 transceiver chip.

Data and signals are received by UART on the Core. UART transfers this information to control circuit. Control circuit sends parameters introduced in the mathematical model to parameter register and initial values of variables of the network image to CNPN Cache. After operating commands are received by CNPN Circuit, the parameters and initial values are transferred to CNPN. This data is processed regarding mathematical model. While iterations are in progress, CNPN Cache sends instantaneous network image to a VGA monitor for real-time observation.

Table 4.2 represents the resource utilization and latencies of arithmetic circuits that are implemented in the Core. Table 4.3 represents the resource utilization of top-level circuits that are designed in the Core.

Table 4.2 Resource utilizations and latencies of arithmetic circuits

Arithmetic circuits	Used slices	Used FFs	Used LUTs	Maximum latency
Fixed-point NPE adder	9	1	18	2
Fixed-point NPE multiplier	9	1	17	2
Fixed-point NPE circuit	395	416	705	50

Table 4.3 Resource utilizations of top-level design

Components of the core	Used slices	Used FFs	Used LUTs
Fixed-point CNPN circuit	6041	6952	10,552
Fixed-point core design	8825	9856	14,295

(a)   **(b)**

(c)   **(d)**

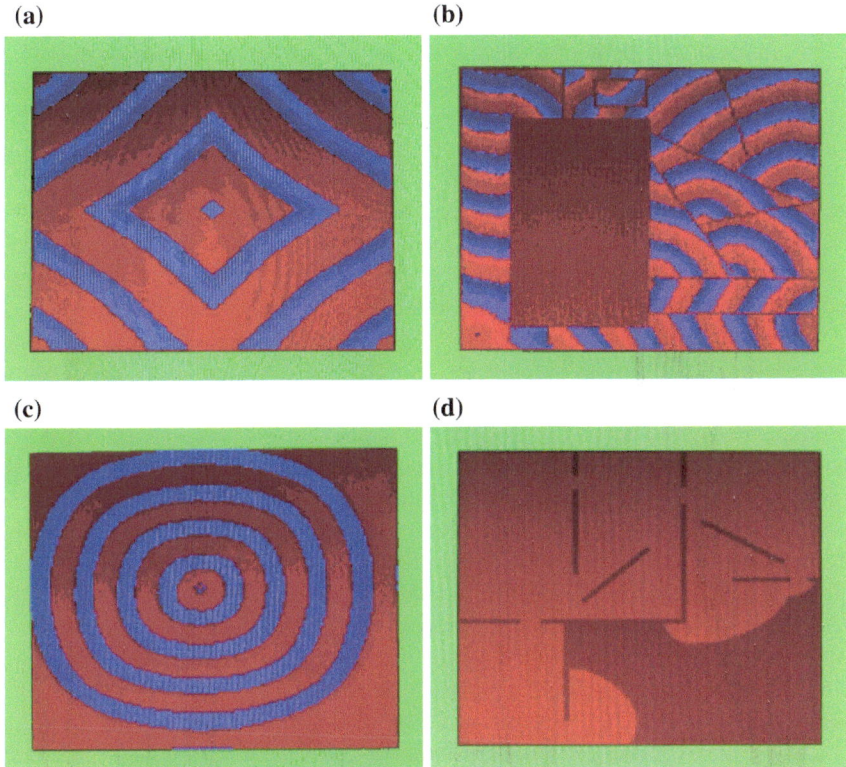

Fig. 4.4 Autowaves and traveling wave are obtained on the network which is emulated by the Core by using different network configurations and initial conditions: **a** autowaves generated from corners, **b** autowaves in medium with obstacles, **c** autowave generated by the center cell, and **d** traveling wave in medium with obstacles

In Fig. 4.4, example network images, which are depicted and emulated by the Core are represented. Autowave illustrations in Fig. 4.4a–c use different initial conditions and network configurations. Figure 4.4d illustrates a traveling wave propagation.

4.2.2 Cellular Nonlinear Network-Based Motion Planning

Motion planning, which has applications from robotics, biochemistry [2], video animations, artificial intelligence [3], to autonomous vehicle navigation [4], is the name of producing a plan that moves object from an initial configuration to a desired configuration while obeying the movement constraint [5]. A state (or configuration) space, either discrete or continuous, which includes all possible values of the variables, such as positions, orientations, velocities of objects like robots, targets, obstacles, is the first fundamental requirement of motion planning. Time is the following requirement. Implicit time definition (having just the order of event sequence) or explicit time definition (having quantities as functions of time) should be provided. State transforming operators, which are called actions, are also required in order to compose a plan to evolve an initial state to the desired one. Henceforth, two criteria are considered to evaluate any algorithm if it is a motion planner. The first one, feasibility, is the indication of success to arrive at the goal state, without efficiency consideration. The second is optimality, which is the indication of a feasible plan which optimizes performance in a clearly specified way. In general, the effort for proposing a feasible solution is more than for an optimal solution in robotics and related fields [5].

In today's applications, sampling-based motion planning, one of the two main approaches in continuous state spaces, is much more referred than combinatorial motion planning, due to its short running times and implementation simplicity [5]. Both methods need geometric modeling of application's world and associated geometric transformations. On the other hand, without geometric representation, the simplest planning algorithms on discrete state space lie at the base of many complex methods or inspire them. Simplicity arises from not only the lack of geometric representations but also the lack of support for differential equations and uncertainty. To build such a motion planner, every unique situation of the world is mapped to a discrete state. The set constituted by those discrete states is called the state space. A state transition function is created, whose inputs are the current state and the action, and output is the next state. A search algorithm which is capable of recording the state transitions is proposed such that the result is a sequence of actions that draws a feasible plan [5]. For an optimal solution, a cost function is defined for actions. Then, the algorithm is enhanced in order to seek for the minimum cost of plan, for example, Dijkstra's algorithm and A* algorithm [6].

The continuous paths generated by both sampling-based and combinatorial motion planning require a feedback controller in real-world applications, because errors and deviations are taken into account. Discrete space motion planning steps forward when embedding such a feedback control to the core of the planner. That planner is called feedback motion planner and produces a feedback plan that involves feasible paths avoiding obstacles by giving an action for every single state. Therefore, any unpredictable deviation in the state of the real world that the planner interacts can be healed by the feedback plan [5]. Potential function Φ, which is a function from the discrete state space to $[0, \infty]$ can be called a navigation function if it satisfies three conditions as follows. (1) $\Phi(x) = 0$ for all x in the goal states set, (2) $\Phi(x) = \infty$ if and only if no point in the goal set is reachable from x, (3) and the local operator gives a next state whose potential is less than the current state for every state excluded the goal set. The local operator may be a minimization operator like the negative gradient operator in continuous space. Navigation function defines a feedback plan if the action is determined by the local operator [5].

Special Cellular Nonlinear Networks (CNNs) serve as feedback motion planners, in the case of $\mathbb{R}^2$ state space, where the states represent the discrete positions on a 2-D Euclidean plane, not the velocity or acceleration, and the only action defined on this state space is 2-D translation of a point object. CNN promises a computation style beyond Boolean logic [7], which is handled by wave computers [8], while algorithmic researches develop algorithms that run sequentially on Boolean processor or runs in parallel on reconfigurable logic [9]. Wave front propagation algorithm, maximum clearance algorithm, and Dial's algorithm in literature yield optimal feedback plans [5]. This phenomenon is also observed in propagation of nonlinear waves, which is also called active waves and spatiotemporal waves. Three different types of wave propagating nonlinear grid networks, which are also called active media, are defined as follows. The first one is excitable networks whose elements (cells) have one stable equilibrium point. Any excitements from outside and from the coupled cells bring the cell out of stable equilibrium, then the cell evolves back to the initial stable state, while the excitation wave is propagating on the network. The second one is bistable networks whose cells have two stable equilibrium points. Excitement brings the cell out of a stable state and evolves to the other stable state. The product wave is called traveling wave or switching wave. The last one is self-oscillatory networks with cells without any stable equilibrium point. The cells typically have a limit cycle in phase portraits causing a periodic oscillation. In this type of networks, cells usually synchronize each other with a proper coupling scheme, and a spatiotemporal event, called autowave, propagates on the network. CNN can properly represent those systems and propagate active waves [10]. Systems capable of propagating binary traveling waves (triggering waves) are also suitable for feedback motion planning [11].

The wave propagation directly generates the feedback motion plan or the navigation function. CNN could be a functional tool for planning. An early work has been published in 1993 that declares a two-dimensional grid array of Chua's circuit is capable of finding an optimal path which needs the least energy even the ground level is wrinkled using different coupling resistors [12]. A simple CNN-based wave propagation algorithm, which has been proposed for real-time robot control, works as a backward search algorithm in which the solution starts from the target point of the searched path [13]. The continuation of that work demonstrates a gathering application of multiple robots [14], which is an easy problem after producing the feedback plan using wave front propagation. Another CNN which is capable of contracting autowaves as well as propagating them is proposed in [15]. This network does not generate feedback plan as the cells do not have memory to save the wave propagation vector. Instead of this, the propagated wave is contracted with two fixed end points. In 2010, the similar results from a FitzHugh–Nagumo network have been announced by Vazquez-Otero [16]. Not only electronic implementations but also chemical setups coupled with electronically implemented active mediums are researched for solving the shortest path problem. In [17, 18], reaction–diffusion mediums are realized by chemical processors with collaborating Cellular Automata (CA) and CNN, respectively. A Cellular Logic Network (CLN) for binary traveling (trigger) wave propagation is designed and applied for morphological image processing in [19, 20]. The CLN is also proper for motion planning [21]. Moreover, the architecture of CLN and CA resemble each other [22].

4.2.3 Generating Feedback Plan by Locally Coupled Oscillatory Network

The fundamental questions are if there is any path for a robot that takes it to the target point, and which one is the shortest if there is more than one path. Wave accumulation and gradient-based algorithm is presented in this subsection to solve these problems. This algorithm produces shorter and smoother paths than wave front diffusion-based algorithm [23] which is the earlier version of the algorithm. In Fig. 4.5 a reference map is presented which will be used to describe and test the algorithms.

Traveling wave front itself offers a solution for the path planning problem. To use this solution, the history of propagating wave can be recorded and processed. Ito et al. [24] expressed a method based on this phenomenon in 2006. For wave accumulation and gradient-based algorithm, all y-state and u-input values are set to 0 s. Initial x-state matrix again has three different nodes: fixed nodes, active non-source nodes and an active source node as shown on reference map in Fig. 4.5. For this algorithm, each node accumulates the values of its x-state during whole emulation and stores it on y-state. To achieve this, the state coefficients are set to $\alpha = 0$, $\beta = 0$, $\epsilon = 1$ and $\sigma = 0$. The weights of the synaptic law are $a_{i,j+1} = 1$, $a_{i-1,j} = 1$, $a_{i,j-1} = 1$ and $a_{i+1,j} = 1$. The slope of the $g(\cdot)$ function is $m = -20$, and its limit value is $limit = 1$. Source node gets the initial value $x_{source} = -1.122$ and other active nodes get the value $x_{initial} = 0$. Fixed boundary and obstacle nodes are set to $x_{fixed} = 0.00005$ at initialization. With this initial value, the network is emulated for 750 iterations. This iteration number gives sufficient time to the wave while covering the whole network.

As shown in Fig. 4.6, traveling wave is generated by the node $(60, 60)$ and it propagates on the x-states. The y-states are the integral of the x-states as presented in Eq. (2.17). At each iteration, the sum of current x-state values and current y-states values are stored on the y-states. Because the initial value of the source node is negative, the y-state of this node becomes smaller at each iteration. Also, this makes the neighbors become smaller but they do not exceed the source node. This rule is valid for all nodes. y-states decrease for all nodes, but none of them exceeds its neighbor which the wave is propagated by. So, the source node always has the minimum y-state value. When the wave covers the whole network, y-state matrix becomes a topographical map. The lowest point on this map is the source node. The highest points are the obstacles and boundaries. Then, the node which is touched last by the wave is the second highest point on the map. Figure 4.7 shows the wave propagation and three-dimensional topographical map evolution step by step. In this configuration, cell's of the network behaves as 1st order stable systems, y state equation is removed from the cell dynamics, and it is employed for just integrating the x state.

After the propagation ends, the coefficients and weights are updated and gradients of this topographical map are computed. To do this, four final iterations are executed. The first final iteration copies the y-states onto x-states, the second one computes only the horizontal gradients of the map, the third one swaps y-states and x-states, and the last one computes only the vertical gradients of the topographical map. To

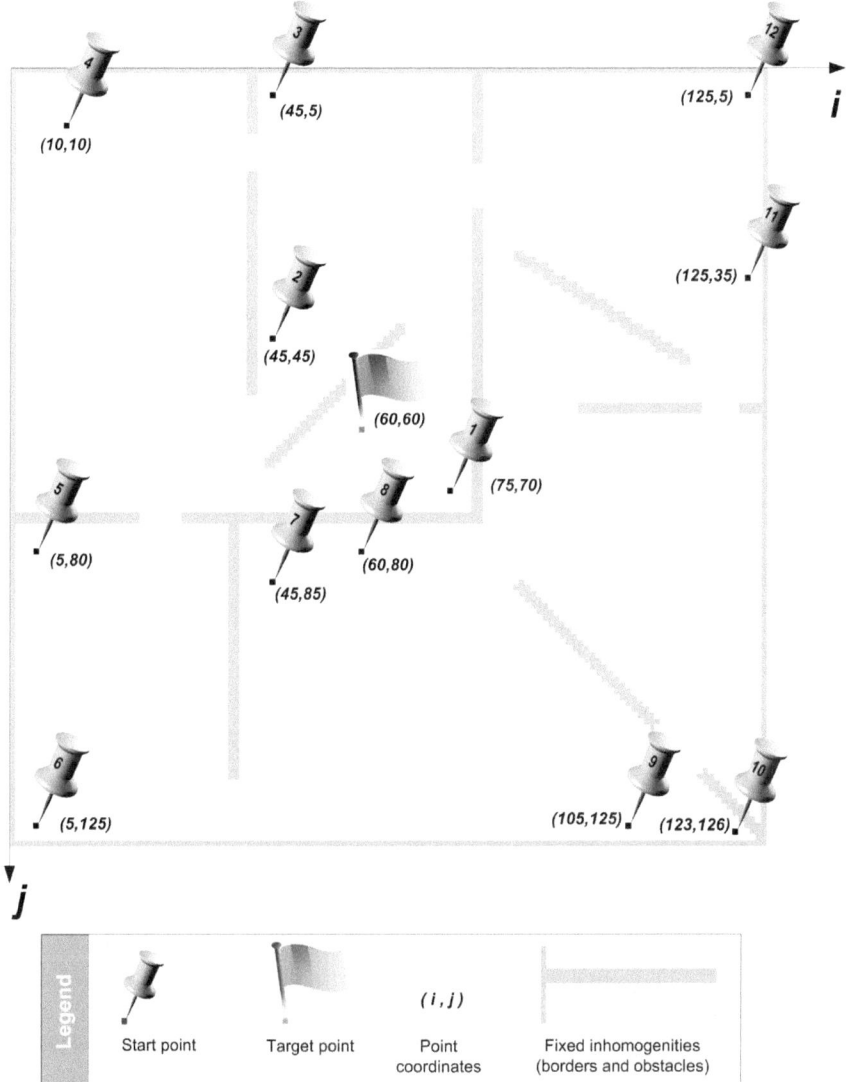

Fig. 4.5 Reference map to test algorithms

execute the first final iteration the parameters are set to $\alpha = -1, \beta = 1, \epsilon = 0, \sigma = 0$, $a_{i,j+1} = 0$, $a_{i-1,j} = 0$, $a_{i,j-1} = 0$, $a_{i+1,j} = 0$, and $m = 0$. The horizontal gradient is computed at the second final step by using the same parameters except $\beta = 0$, $a_{i-1,j} = -1$, and $a_{i+1,j} = 1$. The third iteration swaps the states by using the same parameters except $\beta = 1$, $\epsilon = 1$, $\sigma = -1$, $a_{i,j+1} = 0$, $a_{i-1,j} = 0$, $a_{i,j-1} = 0$, and $a_{i+1,j} = 0$. The last iteration which computes the vertical gradients is executed after

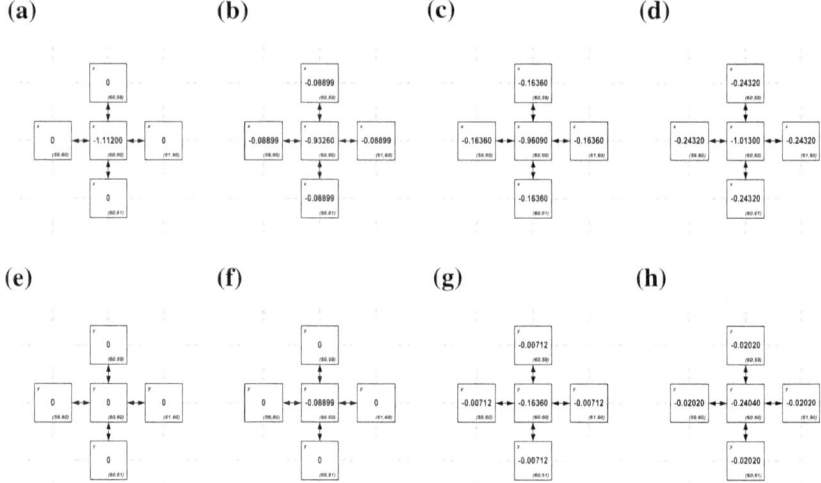

Fig. 4.6 x and y state values of the nodes surrounding the source node during the first three iterations: **a** initial x-states, **b** x-states at the first iteration, **c** x-states at the second iteration, **d** x-states at the third iteration, **e** initial y-states, **f** y-states at the first iteration, **g** y-states at the second iteration, and **h** y-states at the third iteration

the parameters are set to $\beta = 0$, $\epsilon = 0$, $\sigma = 0$, $a_{i,j-1} = -1$, and $a_{i,j+1} = 1$ while others remain the same.

The horizontal and vertical gradient components, which are computed for any active node, are used to draw the path for this algorithm. Following these vectors from starting point guides the robot to target point. Also, giving the initial value $x_{\text{fixed}} = 0.00005$ to the boundaries and obstacles makes a positive accumulation effect to their neighbors and the output gradient vectors move away from them. Strong horizontal vectors are seen next to the vertical walls and strong vertical vectors are seen next to the horizontal walls. This makes the robot avoid from obstacles easily.

Reference map in Fig. 4.5 has 12 start points. Both algorithms ignore these points and consider them as white active nodes while the wave propagating. After vectorial data has been produced, the software draws the path from a determined start point to the target point by using these vectors. The paths from the given 12 start points to the same target point are found by wave front diffusion-based algorithm and drawn on the same map. In Fig. 4.8, the paths are found by wave accumulation and gradient-based algorithm drawn on the reference map (Fig. 4.5).

In [25], digital implementation of LCON (Sect. 4.2) and wave accumulation and gradient-based path finding algorithm which is presented in this subsection are merged for real-time robot navigation problem in dynamically changing environment and presents a real test setup with a roving robot. The setup in [25] has a ceiling camera which captures an image of the top view of the platform at almost every one second. After preprocessing of the captured image, it is loaded to the network which

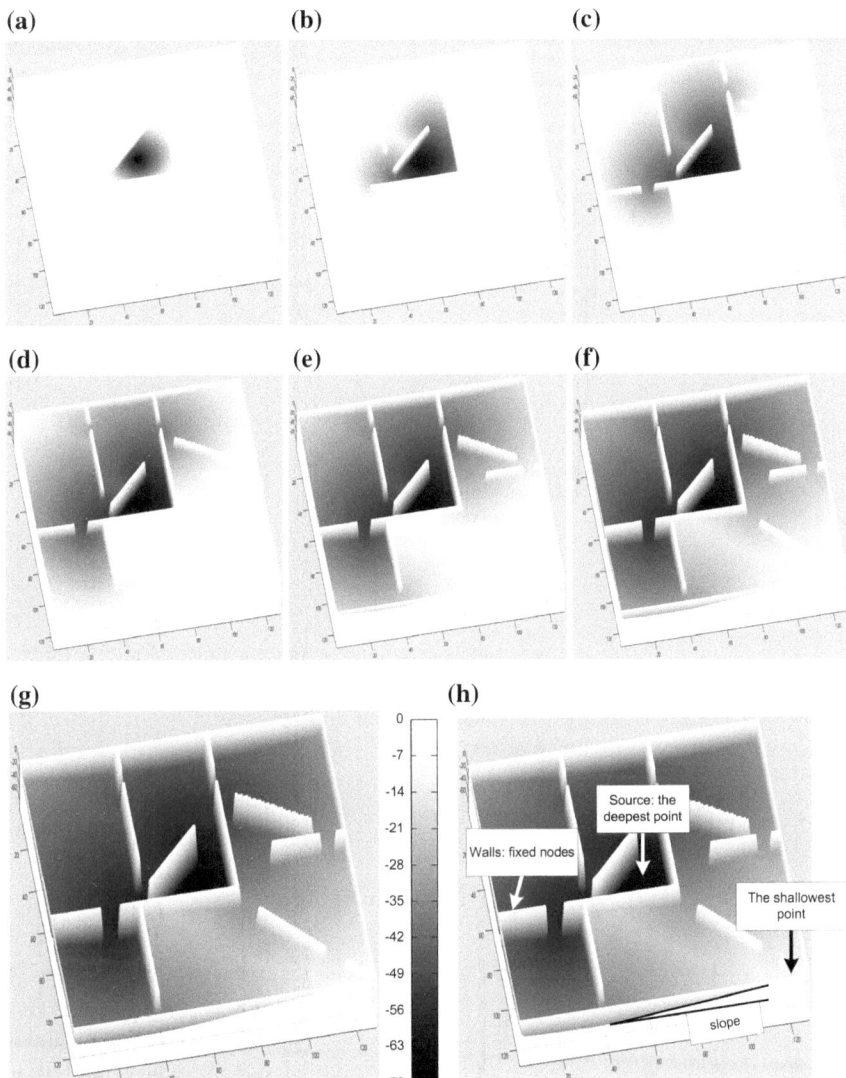

Fig. 4.7 Accumulation images of wave accumulation and gradient-based algorithm with sequencing iterations numbers: **a–g** from top view, and **h** step 700 detailed with labels

is implemented on FPGA and an active wave which is generated by the network is propagated until the robot is covered by the wave. The wave front diffusion-based path finding algorithm always generates a solution for the current state of the platform without the information about its past states. At every repetition of the algorithm, the target position is updated by the system if it is changed.

Fig. 4.8 Traces found by the wave accumulation and gradient-based algorithm

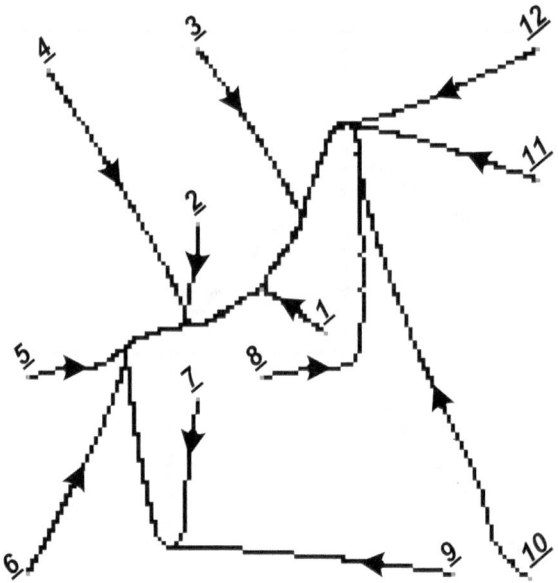

Spatiotemporal wave propagation and Doppler effect on locally coupled oscil-latory network are combined for path planning and navigation application in [26] which gives an opportunity to adjust the trackers speed or change the route com-pletely, dependent to the target motion.

4.3   CNN-Based Artificial Olfaction System

The olfaction systems, in order to classify the odor or give a decision, process the spatiotemporal patterns which are generated by a preprocessing part of the system. The spatiotemporal coding accelerates the decision mechanism, so it compensates the slow response of the olfactory sensors. The similar situation is also valid for the artificial olfaction systems, the response of a widely used commercial odor sensor reaches a stable state in more than a minute. This fact emphasizes the necessity of using the feature extractor or preprocessing block.

In order to imitate the success of the olfaction systems, the collaboration of dif-ferent types of cells that are randomly located in Wilson–Cowan neural mass model should be first imitated in feature extraction part of the system. In Chap. 2, cellu-lar neural network is proposed for implementation of Wilson–Cowan neural mass model and different CNN architectures presented. Reconfigurable cellular neural network is the best for electronic implementation because of the suitability of ran-domly locate the cells in the network. In this section, an artificial olfaction system and it's implementation will be presented.

The artificial olfaction system which is shown in Fig. 4.9 has three main parts: sensing, feature extraction, and processing which are already presented in Sect. 3.3. The sensing part includes an array of 16 metal oxide gas sensors of Figaro Inc.®. The sensor activation is converted into a signal by a simple electrical circuit on electronic board. In this section, the same sensors and measurement setup which was given in Sect. 3.5 is used.

Reconfigurable cellular neural network which is given in Sect. 2.6 works as a feature extractor that maps the spatial code coming from the gas sensors' signal in the sensing part to spatiotemporal patterns. Each cell in the RCNN is thought as a processor and the cells from different subpopulations can be generated from the same type of processors only by an additional variable in the synaptic law because the difference between the subpopulations can be verified by altering a sign. An FPGA implementation of RCNN for feature extraction and its architecture are explained in the following section.

MicroBlaze in FPGA has the flexibility to be programmed as a higher processing unit; however, it is not considered here. The input needed for classifier at the back end can be sent to a Personal Computer (PC) or any other platform. There are various classifiers (such as principal component analysis (Sect. 3.7.1) and support vector machine (Sect. 3.7.2)) running on PC or other platforms, so the classification compartment of the system is not considered in this section.

4.4 Implementation of Processor Population

All the blocks marked with capital P in Fig. 4.9 are individual processors (which realize excitatory and inhibitory neurons) of the processor population. The control circuit has registers which are written by the processor population and by MicroBlaze which is a soft microprocessor core designed for Xilinx FPGAs from Xilinx® [27]. The controller always lets the current iteration step number of the population to MicroBlaze. MicroBlaze can also set a termination limit to the iteration, and controller ensures the obedience of the population.

The identity of P processors (which is defined by ID matrix H given in Sect. 2.6) is driven by this control circuit, but the basic source of their identity is MicroBlaze. The block diagram of an individual processor P is shown in Fig. 4.10. In the current state of our work, the sensory data inputs of the P processors are provided by MicroBlaze. MicroBlaze is able to send the inputs, initial states and receive the computed states of the P processors through this control circuit.

IGEN and FEI are subblocks of a P processor in RCNN. The synaptic law $\mathfrak{J}_{i,j}$ which is given in Eq. (2.35) is computed by IGEN subblock. The state equation which is given in Eq. (2.36) is computed by FEI subblock. These two subblocks have their own datapath and controller.

Sensory input signal $U_{i,j}[k]$, cell ID $h_{i,j}$ (or P processor ID), current state values at the corresponding coordinate (i, j) and neighbors' current states values together with identities in its *sphere of influence* are the inputs of datapath of IGEN subblock.

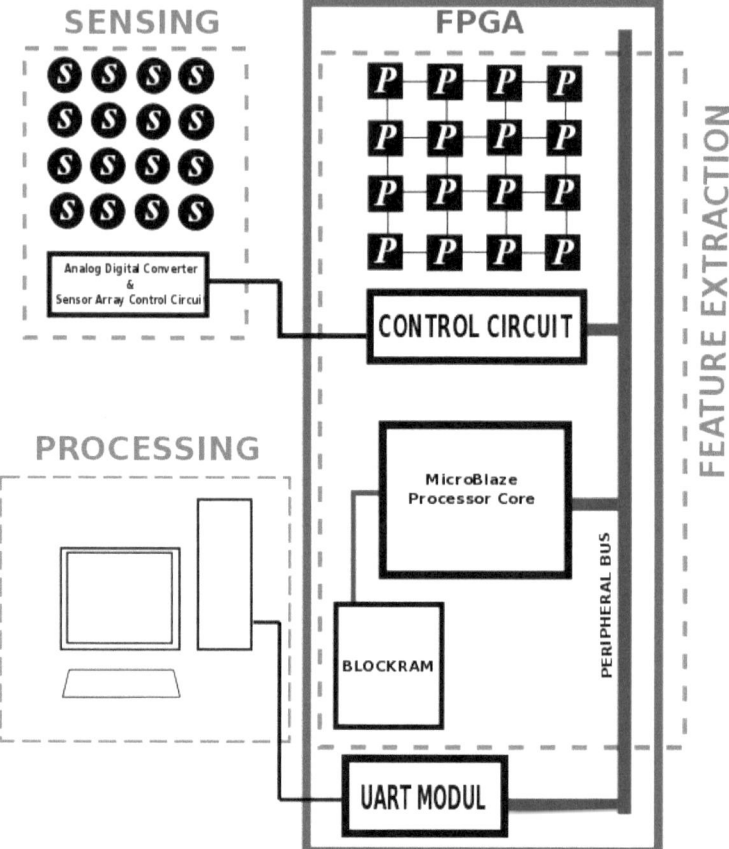

Fig. 4.9 Reconfigurable cellular neural network consists of 16 processor cells P which is the feature extraction part maps the sensor information to the spatial code. The sensing part includes an array of 16 metal oxide gas sensors of Figaro Inc.®. Processor population is controlled by an 32-bit general-purpose processor so-called MicroBlaze IP Core [27]. MicroBlaze's peripheral bus is used to connect the blocks. MicroBlaze has UART connection with outside. MicroBlaze is used to program to the processing unit such as ID matrix. One can also program the MicroBlaze to perform processing part of artificial olfaction system for decision such as classification (© 2019 IEEE. Reprinted, with permission, from Ayhan et al. [28])

This subblock serves to compute synaptic law (2.35) for kth iteration step. The next state value $x_{i,j}[k+1]$ is computed in the datapath of FEI subblock which uses $\Im_{i,j}[k]$ output from IGEN and its own current state with two weights α and β. Controllers of IGEN and FEI subblocks are connected with IGEN_done and FEI_ack signals, because they run consecutively.

The controller of IGEN is a three-state Finite-State Machine (FSM). In the idle state, the start signal is being waited to be logical high. When it is asserted, IGEN goes to a blank state, because its datapath has more propagation delay than the

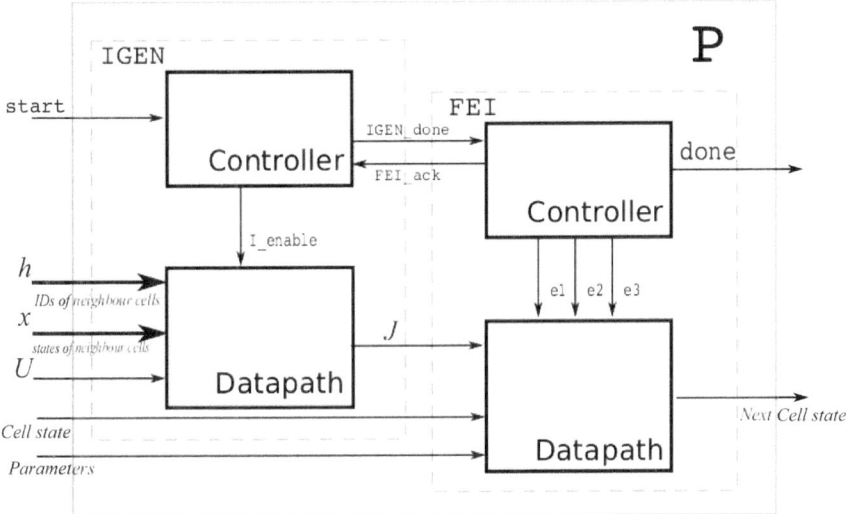

Fig. 4.10 The block diagram of an individual processor P

system's master clock period. After IGEN $\mathfrak{I}_{i,j}[k]$ is computed, the controller asserts done signal for the controller of FEI. $S_{i,j}(1)$ which is defined in (2.10)

The datapath of IGEN subblock is shown in Fig. 4.11, IGEN computes $\mathfrak{I}_{i,j}[k]$ by

$$\mathfrak{I}_{i,j}[k] = \begin{cases} J[k] \text{ if } J[k] > 0; \\ 0 \quad \text{ else}; \end{cases} \tag{4.5}$$

and

$$J[k] = \frac{1}{2}\Big\{(-1)^{h_{i,j}}\big[(h_{i,j} \oplus h_{i,j-1})x_{i,j-1}[k] \\ + (h_{i,j} \oplus h_{i-1,j})x_{i-1,j}[k] + (h_{i,j} \oplus h_{i+1,j})x_{i+1,j}[k] \tag{4.6} \\ + (h_{i,j} \oplus h_{i,j+1})x_{i,j+1}[k]\big] + U_{i,j}[k]\Big\},$$

where $\oplus$ denotes logic EX-OR operation which helps to connect only the neurons belonging to different subpopulations. As it is seen in Fig. 4.11, the outputs of EX-OR operations select the MUXes' inputs and the adder has a 0 or the neighbor's state. The sum is complemented (inhibitory) or not (excitatory) depend on the cell $C_{i,j}$ ID ($h_{i,j}$). If the cell belongs to the inhibitor subpopulation, then the sum is added to the $U_{i,j}[k]$ input. Otherwise, the sum is subtracted from the $U_{i,j}[k]$ input. The value calculated is compared by 0 because of the activation function θ in Eq. (2.35) which is heaviside step function. After the comparison, synaptic law (2.35) for kth iteration step ($\mathfrak{I}_{i,j}[k]$) directly saved to the output register. On the other hand, if not, the output register saves 0.

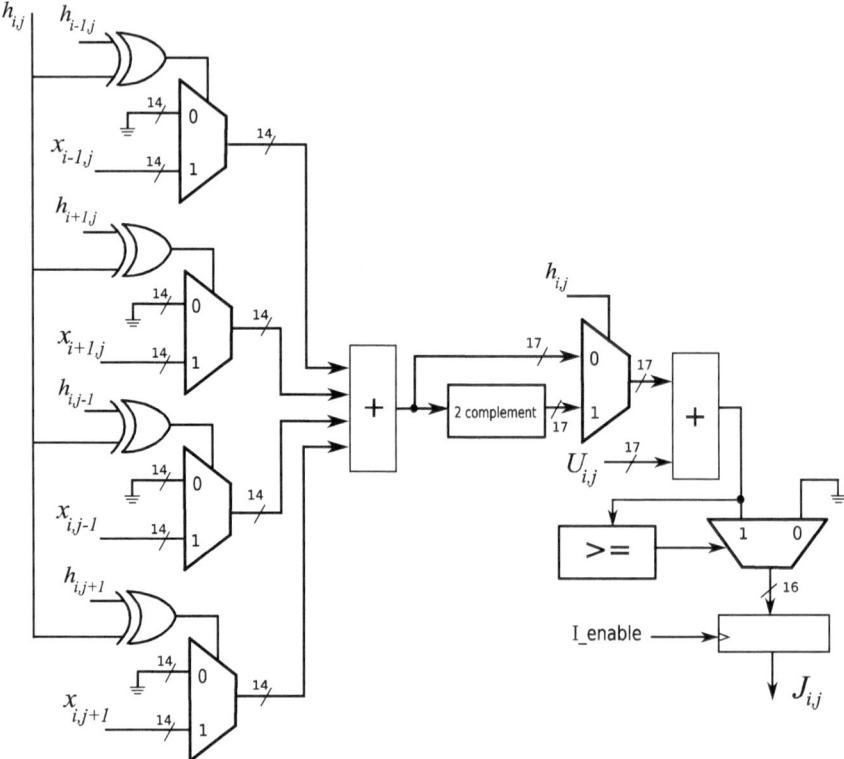

Fig. 4.11 The datapath of IGEN subblock

The finite-state machine of the FEI subblock has four states. When the controller of IGEN passes the activation by asserting its done output (IGEN_done), the datapath of FEI calculates $\frac{1}{2^{16}} \times \alpha \times \Im_{i,j}[k]$ and stores the result into Reg_1. While IGEN is computing, FEI calculates $\frac{1}{2^{16}} \times \beta \times x_{i,j}[k]$ and stores into Reg_2. By the following state, these stored products and $x_{i,j}[k]$ are accumulated. The result is captured by output register Reg_3 of FEI and appears on the $x_{i,j}[k+1]$ output. While returning to the idle state, FEI outputs a done pulse, which is sensed by the hybrid processor population control circuit.

Figure 4.12 shows two MUXes, and a DEMUX logic block prepares a basis to realize two multiplication using one multiplier. The starting signal controls the FEI block. The structure is not a different one than two cascade-connected adders. The output register of the datapath of FEI is controlled by its finite-state machine.

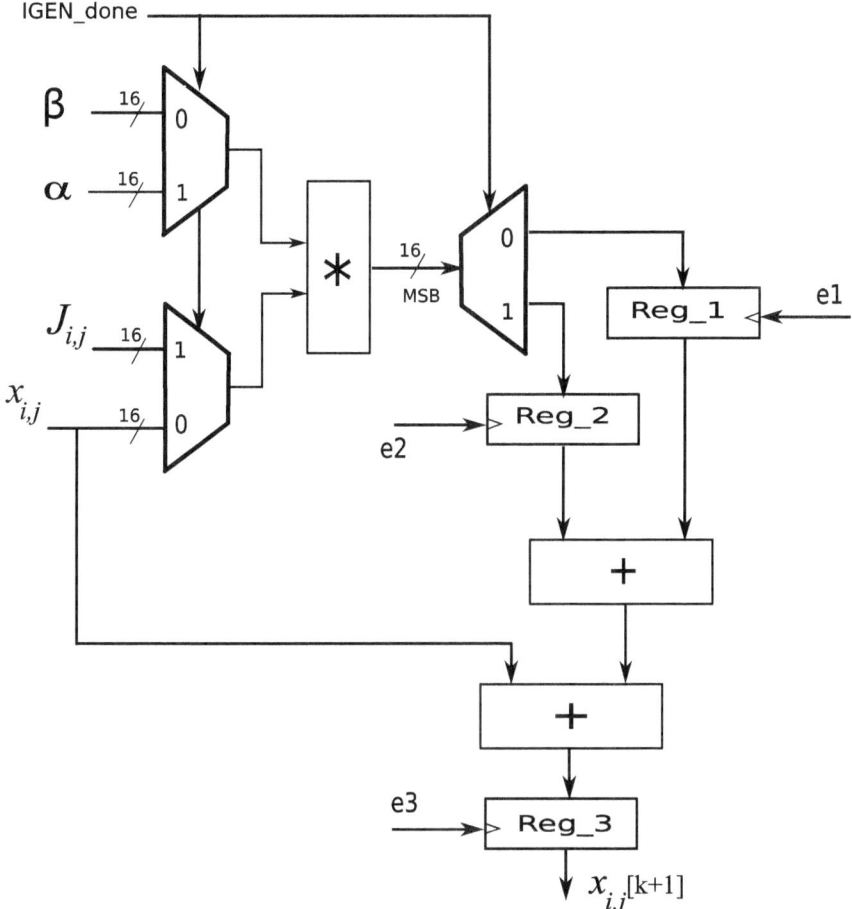

Fig. 4.12 The datapath of FEI subblock

4.4.1 Implementation Results

Reconfigurable cellular neural network consists of 16 processor cells which is given
in Fig. 4.9 was implemented on Spartan-3E 1600 FPGA consumes 34% of the logical
slices [28]. The system also uses 11% of BlockRAM resources and 52% of 18 ×
18 dedicated multiplier blocks. When the processor population is tried to expand
from 4 × 4 to 5 × 5 P processors the usage of slices becomes 46% and usage of
dedicated multipliers becomes 77% of the chip. BlockRAM usage changes when a
more complicated software will be run on the MicroBlaze Core.

The implementation [28] on FPGA is fully synchronous and uses 20 ns clock
signal. A single forward Euler integration step occurs on processor population in

140 ns using 16-bit fixed-point number representation. This proposed system is also allowed to emulate larger than 4×4 processors population.

Local binary descriptor and local binary pattern operations are the main processing blocks in the smart camera implementations. These operations calculate one bit of the descriptor for each pixel in an image comparing the pixel with the same pair of pixels or combinations of pixels. The diversity of comparisons and the diversity of network topology are linearly depend on each other. However, reconfigurable cellular neural networks provide multiple network topologies in the same architecture without any additional hardware cost.

Hybrid processor population network which is also designed based on reconfigurable cellular neural network architecture was introduced and implemented for local binary feature description applications by Ergunay and Leblebici [29]. In [30], run time programmability due to the reconfigurability of the network is utilized in object detection applications where multiple types of targets exist in the scene.

References

1. R. Yeniceri, M.E. Yalcin, An emulated digital wave computer core implementation, in *European Conference on Circuit Theory and Design* (2009), pp. 831–834
2. J.C. Latombe, Motion planning: a journey of robots, molecules, digital actors, and other artifacts. Int. J. Robot. Res. **18**(11), 1119–1128 (1999)
3. S. LaValle, Motion planning: robotics automation magazine. IEEE **18**(1), 79–89 (2011)
4. E. Koyuncu, N. Ure, G. Inalhan, Integration of path/maneuver planning in complex environments for agile maneuvering UCAVs. J. Intel. Robot. Syst. **57**(1–4), 143–170 (2010)
5. S.M. LaValle, *Planning Algorithms* (Cambridge University Press, New York USA, 2006)
6. T.H. Cormen, C.E. Leiserson, R.L. Rivest, C. Stein, *Introduction to Algorithms* (The MIT Press, Cambridge, Massachusetts, 2009)
7. T. Roska, Circuits, computers, and beyond boolean logic. Int. J. Circuit Theory Appl. **35**(5–6), 485–496 (2007)
8. T. Roska, Cellular wave computers for nano-tera-scale technology—beyond Boolean, spatial-temporal logic in million processor devices. Electron. Lett. **43**(8), 427–429 (2007)
9. N. Atay, B. Bayazit, A motion planning processor on reconfigurable hardware, in *2006 Proceedings of the IEEE International Conference on Robotics and Automation, ICRA* (2006), pp. 125–132
10. L.O. Chua, M. Hasler, G.S. Moschytz, J. Neirynck, Autonomous cellular neural networks: a unified paradigm for pattern formation and active wave propagation. IEEE Trans. Circ. Syst. I Fundam. Theory Appl. **42**(10), 559–577 (1995)
11. C. Rekeczky, L.O. Chua, Computing with front propagation: active contour and skeleton models in continuous-time CNN. J. VLSI Signal Process. Syst. **23**(2/3), 373–402 (1999)
12. V. Perez-Munuzuri, V. Perez-Villar, L.O. Chua, Autowaves for image processing on a two-dimensional CNN array of excitable nonlinear circuits: flat and wrinkled labyrinths. IEEE Trans. Circuits Syst. I Fundam. Theory Appl. **40**(3), 174–181 (1993)
13. A. Gacsadi, T. Maghiar, V. Tiponut, A CNN path planning for a mobile robot in an environment with obstacles, in *2002 Proceedings of the 7th IEEE International Workshop on Cellular Neural Networks and Their Applications, (CNNA 2002)* (2002), pp. 188–194
14. I. Gavrilut, V. Tiponut, A. Gacsadi, C. Grava, CNN processing techniques for multi-robot coordination, in *2007 International Symposium on Signals, Circuits and Systems, ISSCS* (2007), pp. 1–4

15. A.P. Munuzuri, A. Vazquez-Otero, The CNN solution to the shortest-path-finder problem, in *2008 11th International Workshop on Cellular Neural Networks and Their Applications* (2008), pp. 248–251

16. A. Vazquez-Otero, A.P. Munuzuri, Navigation algorithm for autonomous devices based on biological waves, in *2010 12th International Workshop on Cellular Nanoscale Networks and Their Applications (CNNA)* (2010), pp. 1–5

17. A. Adamatzky, B.D. Costello, Reaction-diffusion path planning in a hybrid chemical and cellular-automaton processor. Chaos Solitons Fractals **16**(5), 727–736 (2003)

18. A. Adamatzky, P. Arena, A. Basile, R. Carmona-Galan, B.D.L. Costello, L. Fortuna, M. Frasca, A. Rodriguez-Vazquez, Reaction-diffusion navigation robot control: from chemical to VLSI analogic processors. IEEE Trans. Circ. Syst. I Regul. Pap. **51**(5), 926–938 (2004)

19. P. Dudek, An asynchronous cellular logic network for trigger-wave image processing on fine-grain massively parallel arrays. IEEE Trans. Circ. Syst. II Express Briefs **53**(5), 354358 (2006)

20. A. Lopich, P. Dudek, Asynchronous cellular logic network as a co-processor for a general-purpose massively parallel array. Int. J. Circuit Theory Appl. **39**(9), 963972 (2011)

21. R. Yeniceri, E. Abtioglu, B. Govem, M.E. Yalcin, A 1616 Cellular logical network with partial reconfiguration feature, in *2014 14th International Workshop on Cellular Nanoscale Networks and their Applications (CNNA)* (2014), pp. 1–2

22. P.G. Tzionas, A. Thanailakis, P.G. Tsalides, Collision-free path planning for a diamond-shaped robot using two-dimensional cellular automata. IEEE Trans. Robot. Autom. **13**(2), 237–250 (1997)

23. R. Yeniceri, M.E. Yalcin, Path planning on cellular nonlinear network using active wave computing technique. Bioeng. Bioinspired Syst. IV **7365**(1), 736508 (2009)

24. K. Ito, M. Hiratsuka, T. Aoki, T. Higuchi, A shortest path search algorithm using an excitable digital reaction-diffusion system. IEICE Trans. Fundam. Electron. Commun. Comput. Sci. **E89-A**(3), 735–743 (2006)

25. V. Kilic, R. Yeniceri, M.E. Yalcin, A new active wave computing based real time mobile robot navigation algorithm for dynamic environment, in *2010 12th International Workshop on Cellular Nanoscale Networks and Their Applications (CNNA)* (2010), pp. 1–6

26. R. Yeniceri, M.E. Yalcin, A new CNN based path planning algorithm improved by the Doppler effect, in *2012 13th International Workshop on Cellular Nanoscale Networks and Their Applications (CNNA)* (2012), pp. 1–5

27. MicroBlaze, https://www.xilinx.com/products/design-tools/microblaze.html Date of access: February 15, 2019

28. T. Ayhan, R. Yeniceri, S. Ergunay, M.E. Yalcin, Hybrid processor population for odor processing, in *IEEE International Symposium on Circuits and Systems (ISCAS)* (2012), pp. 177–180

29. S. Ergunay, Y. Leblebici, Hardware implementation of a smart camera with keypoint detection and description, in *IEEE International Symposium on Circuits and Systems (ISCAS)* (2018), pp. 1–4

30. S. Ergunay, A smart camera architecture for wireless and multiple camera applications. Doctoral dissertation, Ecole Polytechnique Feederale de Lausanne (2018)

Index

© The Author(s), under exclusive licence to Springer Nature Switzerland AG 2020   73
M. E. Yalçın et al., *Reconfigurable Cellular Neural Networks and Their Applications*,
SpringerBriefs in Nonlinear Circuits, https://doi.org/10.1007/978-3-030-17840-6